T0332247

Mathematical Models in Electrical Circuits:
Theory and Applications

Mathematics and Its Applications

Volume 66

Mathematical Models in Electrical Circuits: Theory and Applications

by

C. A. Marinov and
P. Neittaanmäki
Department of Mathematics,
University of Jyväskylä,
Jyväskylä, Finland

KLUWER ACADEMIC PUBLISHERS
DORDRECHT / BOSTON / LONDON

Library of Congress Cataloging-in-Publication Data

Marinov, C. A. (Corneliu A.), 1947-
   Mathematical models in electrical circuits : theory and
applications / by C.A. Marinov and P. Neittaanmaki.
      p.   cm. -- (Mathematics and its applications ; v. 66)
   Includes bibliographical references (p.   ) and index.
   ISBN 0-7923-1155-8 (HB : alk. paper)
   1. Electric circuits--Mathematical models.   I. Neittaanmaki, P.
(Pekka) II. Title. III. Series: Mathematics and its applications
(Kluwer Academic Publishers) ; v.66.
   TK454.M28 1991
   621.319'2'011--dc20   91-25730

ISBN 0-7923-1155-8

Published by Kluwer Academic Publishers,
P.O. Box 17, 3300 AA Dordrecht, The Netherlands.

Kluwer Academic Publishers incorporates
the publishing programmes of
D. Reidel, Martinus Nijhoff, Dr W. Junk and MTP Press.

Sold and distributed in the U.S.A. and Canada
by Kluwer Academic Publishers,
101 Philip Drive, Norwell, MA 02061, U.S.A.

In all other countries, sold and distributed
by Kluwer Academic Publishers Group,
P.O. Box 322, 3300 AH Dordrecht, The Netherlands.

Printed on acid-free paper

SERIES EDITOR'S PREFACE

'Et moi, ..., si j'avait su comment en revenir,
je n'y serais point allé.'

Jules Verne

The series is divergent; therefore we may be
able to do something with it.

O. Heaviside

One service mathematics has rendered the
human race. It has put common sense back
where it belongs, on the topmost shelf next
to the dusty canister labelled 'discarded non-
sense'.

Eric T. Bell

Mathematics is a tool for thought. A highly necessary tool in a world where both feedback and non-linearities abound. Similarly, all kinds of parts of mathematics serve as tools for other parts and for other sciences.

Applying a simple rewriting rule to the quote on the right above one finds such statements as: 'One service topology has rendered mathematical physics ...'; 'One service logic has rendered computer science ...'; 'One service category theory has rendered mathematics ...'. All arguably true. And all statements obtainable this way form part of the raison d'être of this series.

This series, *Mathematics and Its Applications*, started in 1977. Now that over one hundred volumes have appeared it seems opportune to reexamine its scope. At the time I wrote

"Growing specialization and diversification have brought a host of monographs and textbooks on increasingly specialized topics. However, the 'tree' of knowledge of mathematics and related fields does not grow only by putting forth new branches. It also happens, quite often in fact, that branches which were thought to be completely disparate are suddenly seen to be related. Further, the kind and level of sophistication of mathematics applied in various sciences has changed drastically in recent years: measure theory is used (non-trivially) in regional and theoretical economics; algebraic geometry interacts with physics; the Minkowsky lemma, coding theory and the structure of water meet one another in packing and covering theory; quantum fields, crystal defects and mathematical programming profit from homotopy theory; Lie algebras are relevant to filtering; and prediction and electrical engineering can use Stein spaces. And in addition to this there are such new emerging subdisciplines as 'experimental mathematics', 'CFD', 'completely integrable systems', 'chaos, synergetics and large-scale order', which are almost impossible to fit into the existing classification schemes. They draw upon widely dif%ferent sections of mathematics."

By and large, all this still applies today. It is still true that at first sight mathematics seems rather fragmented and that to find, see, and exploit the deeper underlying interrelations more ef%fort is needed and so are books that can help mathematicians and scientists do so. Accordingly MIA will continue to try to make such books available.

If anything, the description I gave in 1977 is now an understatement. To the examples of interaction areas one should add string theory where Riemann surfaces, algebraic geometry, modular functions, knots, quantum field theory, Kac-Moody algebras, monstrous moonshine (and more) all come together. And to the examples of things which can be usefully applied let me add the topic 'finite geometry'; a combination of words which sounds like it might not even exist, let alone be applicable. And yet it is being applied: to statistics via designs, to radar/sonar detection arrays (via finite projective planes), and to bus connections of VLSI chips (via dif%ference sets). There seems to be no part of (so-called pure) mathematics that is not in immediate danger of being applied. And, accordingly, the applied mathematician needs to be aware of much more. Besides analysis and numerics, the traditional workhorses, he may need all kinds of combinatorics, algebra, probability, and so on.

In addition, the applied scientist needs to cope increasingly with the nonlinear world and the extra mathematical sophistication that this requires. For that is where the rewards are. Linear models are honest and a bit sad and depressing: proportional ef%forts and results. It is in the non-linear world that infinitesimal inputs may result in macroscopic outputs (or vice versa). To appreciate what I am hinting at: if electronics were linear we would have no fun with transistors and computers; we would have no TV; in fact you would not be reading these lines.

There is also no safety in ignoring such outlandish things as nonstandard analysis, superspace and anticommuting integration, p-adic and ultrametric space. All three have applications in both electrical engineering and physics. Once, complex numbers were equally outlandish, but they frequently proved the shortest path between 'real' results. Similarly, the first two topics named have already provided a number of 'wormhole' paths. There is no telling where all this is leading - fortunately.

Thus the original scope of the series, which for various (sound) reasons now comprises five sub-series: white (Japan), yellow (China), red (USSR), blue (Eastern Europe), and green (everything else), still applies. It has been enlarged a bit to include books treating of the tools from one subdiscipline which are used in others. Thus the series still aims at books dealing with:

- a central concept which plays an important role in several dif%ferent mathematical and/or scientific specialization areas;
- new applications of the results and ideas from one area of scientific endeavour into another;
- influences which the results, problems and concepts of one field of enquiry have, and have had, on the development of another.

To quote the authors: 'In any mathematical approach to a real world problem, the first step is to establish an adequate and useful mathematical model'. Here 'adequate' and 'useful' means that once these problems are understood (more or less completely) the solutions and techniques are indeed of value in practice. That is indeed the case in the present instance: the modelling of electronic and electrical circuits, notably MOS circuits. Fortunately for the mathematicians, the mathematics involved is nontrivial: and, fortunately for the engineers, techniques are far enough advanced to be able to calculate and design in concrete situations.

Electrical and electronic engineering has an enviable reputation as an area in which the intuitions of engineers and the preciseness and exactness (and sometimes surprising insights) of mathematics reinforce each other rather than fight each other. This book will do much to enhance that reputation.

The shortest path between two truths in the
real domain passes through the complex
domain.

J. Hadamard

La physique ne nous donne pas seulement
l'occasion de résoudre des problèmes ... elle
nous fait pressentir la solution.

H. Poincaré

Never lend books, for no one ever returns
them; the only books I have in my library
are books that other folk have lent me.

Anatole France

The function of an expert is not to be more
right than other people, but to be wrong for
more sophisticated reasons.

David Butler

Amsterdam, August 1991   Michiel Hazewinkel

Contents

Preface

In any mathematical approach to a real-world problem, the first step is to establish an adequate and useful mathematical model. This book comprises of the authors research on mathematical models of electrical and electronic circuits. Especially, we describe applications of the theory of ordinary and partial differential equations to electrical networks. Well-posedness theorems are proved and discussed for nonlinear circuits with lumped parameters, the modelling of bipolar transistors, as well as for linear networks with distributed parameters describing MOS circuits. Special importance is given to the asymptotic behaviour of solutions, with results of practical interest regarding to the operating speed (delay time) of digital integrated circuits.

We have made a constant effort to make the book as self-contained as possible. Of course, familiarity with basic facts in functional analysis, differential equations and circuit theory is assumed.

The material of the book is organized as follows. Chapter I introduces the reader to mathematical tools that are efficient in handling the obtained models. Especially, we give a short introduction to the theory of dissipative (monotone) operators. All circuit models presented in the following chapters lead to differential equations (for the dynamic process) or to time-independent equations (for the steady state) implying a dissipative operator on $\mathbf{R}^n$ (for lumped parameter circuits in Chapter II), on ℓ^p (for "infinite" circuits in Chapter III) or on a space of the form $L^2 \times \mathbf{R}^n$ (for mixed type circuits in Chapters IV, V and VI). Our goal is to prove qualitative properties of these models: the existence and uniqueness of steady state and dynamic solutions in a well precised sense (classical, distribution-type, weak), their boundedness, stability and source dependent continuity. Lumped parameter circuits with nonlinearly modelled bipolar transistors are studied in Chapter II. Chapter III deals with circuits containing an infinite number of lumped parameters where transistors are also present. The last three chapters treat a very large class of circuits in which distributed parameter elements (modelled by degenerate Telegraph Equations) are connected with lumped resistive and capacitive elements. It is an appropriate model ("mixed-type") for studying the influence of interconnections on the global behavior of an integrated circuit with MOS transistors. The proof of asymptotic stability implies an easily computable upper bound of the performance parameter named "delay time", an expression of the speed of signal propagation in the network and an essential quality parameter of digital integrated circuits. (Otherwise, in Chapter II similar bounds were derived.) A numerical procedure to solve mixed-type circuits based on finite elements is given in Chapter VI. Several examples illustrate the method and verify the previously infered bound of delay time.

Very often the mathematicians reproach engineers with the "intuitive" (not rigorous) character of their reasonings, while the engineers are displeased with "academic" (not realistic) problems on which the mathematicians work. Our book proposes an armistice. On the one hand it uses efficient tools developed in functional analysis, differential equations and numerical analysis to solve problems of circuit theory. Conversly, to applied mathematicians, the book offers new models derived from the engineering of integrated circuits, a field in full and fast progress.

We would like to express our graditude to Dr. V. Hara and Dr. A. Lehtonen for their stimulating collaboration and to J-P. Santanen Ph.lic. for his help in preparation of this book. We wish to thank P. Kemppainen M.Sc. and A. Roikonen M.Sc. for their skilful typing of the manuscript in TeX. We are also indebted to the Academy of Finland, Research Council for Technology for their financial support.

Jyväskylä, Finland   C.A. Marinov
May, 1991   P. Neittaanmäki

Chapter I

Dissipative operators and
differential equations on Banach spaces

1.0. Introduction

If we consider the initial value problem

$$x'(t) = f(t, x(t)), \quad x(0) = x_0$$

on the real line, it is well known that one–sided bounds like

$$[f(t, x) - f(t, y)] (x - y) \leq \omega(x - y)^2$$

give much better information about the behaviour of solutions than the Lipschitz–type estimates

$$|f(t, x) - f(t, y)| \leq L|x - y| ,$$

because ω, unlike L, may be negative. This fact was extended to the case when, for all t, the operator $f(t, \cdot)$ has the domain and range in a Hilbert space satisfying

$$\langle f(t, x) - f(t, y), x - y \rangle \leq 0 .$$

Such an operator has been called *dissipative* since the energy of the corresponding system does not increase (Phillips [1957,1959]). However, there are several problems in partial differential equations where it is more natural to consider Banach spaces rather than Hilbert spaces. To simulate the inner product techniques on Banach spaces the concept "semi–inner product" has been introduced and one has studied operators which are dissipative with respect to such a semi–inner product (Lumer and Phillips [1961]).

During the last thirty years the theory of dissipative operators in connection with the theory of semigroups has been proved to be a fertile mathematical field. Brézis [1973,1987], Goldstein [1970], Barbu [1976], Martin [1976], Pazy [1983] are few of the more well-known textbooks of the domain. Let us note that many authors prefer to call A monotone or accretive if $-A$ is dissipative. Various applications

1

in fluid mechanics, chemical processes, biology, medicine, ecology and in economics encouraged the development of the theory.

In view of defining — in Section 1.2 — the dissipative operators on a Banach space, Section 1.1 deals with properties of "duality-type" functionals, their relation with duality operators and with norm subdifferentials and their concrete expressions in L^p and C spaces.

Section 1.3 proves the classical results of Hille–Yosida and Lumer–Phillips regarding the generation of semigroups, while Section 1.4 deals with linear and the quasiautonomous Cauchy problem with generating semigroups terms.

Section 1.5 analyses the Cauchy problem with a nonlinear, dissipative and time dependent operator whose domain is time-independent.

1.1. Duality type functionals

Let X be a Banach space over the field $\mathbf{K}$ ($\mathbf{R}$ or $\mathbf{C}$) and let $\| \cdot \|$ be the norm on X. Strong neighbourhoods of a point $x \in X$ are the open balls $B(x, r) = \{y \in X \; ; \; \|y - x\| < r\}$ for any $r > 0$. The boundary and the closure of $B(x, r)$ are $S(x, r) = \partial B(x, r) = \{y \in X \; ; \; \|y - x\| = r\}$ and $\overline{B}(x, r) = B(x, r) \cup S(x, r)$ respectively.

The space X^* of all linear continuous functionals on X is called the dual space of X. The space X^* is also Banach with respect to the norm

$$\|f\| = \sup \{|f(x)| \; ; \; \|x\| \leq 1\} .$$

Beside the strong topology defined by the norm we shall also consider the weak topology in X. This locally convex topology is given by the family of seminorms $\{p_f(x) = |f(x)| \; ; \; f \in X^*\}$. We shall denote by $\xrightarrow{w}$ the convergence in this topology ("weak convergence").

On X^* we shall consider also the weak*–topology (locally convex and separate) given by the seminorms $\{p_x(f) = |f(x)| \; ; \; x \in X\}$. The corresponding weak*–convergence will be denoted by $\xrightarrow{w^*}$.

Let us take $x, y \in X$ and $\theta \in]0, 1[$. The fact that for all $h > 0$ we have

$$\|x + \theta hy\| = \|\theta(x + hy) + (1 - \theta)x\| \leq \theta \|x + hy\| + (1 - \theta)\|x\|$$

shows us that

$$-\|y\| \leq (\theta h)^{-1} (\|x + \theta hy\| - \|x\|) \leq h^{-1} (\|x + hy\| - \|x\|) \leq \|y\|.$$

As an immediate consequence we find

$$\inf \left\{ \frac{\|x + hy\| - \|x\|}{h} \; ; \; h > 0 \right\} = \lim_{h \to 0+} \frac{\|x + hy\| - \|x\|}{h} =: \langle x, y \rangle_+$$

and

$$\sup\left\{\frac{\|x + hy\| - \|x\|}{h}\ ;\ h < 0\right\} = \lim_{h \to 0^-}\frac{\|x + hy\| - \|x\|}{h} =: \langle x, y\rangle_-\ .$$

We call "duality type functionals" the functions $\langle\ ,\ \rangle_{\pm} : X \times X \mapsto \mathbf{R}$. They play a crucial role in this book. For convenience of future references we list several elementary properties of these functionals. The proofs are rather straightforward and can be found (except xi)) in Martin [1976, Lemma 5.6, Ch. 2].

Lemma 1.1. *If $\langle\ ,\ \rangle_{\pm}$ is defined as above, then:*

    i) $\langle x, y\rangle_+ = -\langle x, -y\rangle_-$
   ii) $\langle x, y_1 + y_2\rangle_+ \le \langle x, y_1\rangle_+ + \langle x, y_2\rangle_+$
  iii) $\langle x, y_1 + y_2\rangle_+ \ge \langle x, y_1\rangle_+ + \langle x, y_2\rangle_-$
  iv) $\langle x, y_1 + y_2\rangle_- \le \langle x, y_1\rangle_+ + \langle x, y_2\rangle_-$
   v) $\langle x, y_1 + y_2\rangle_- \ge \langle x, y_1\rangle_- + \langle x, y_2\rangle_-$
  vi) $-\|y\| \le \langle x, y\rangle_- \le \langle x, y\rangle_+ \le \|y\|$
 vii) $\left|\langle x, y\rangle_{\pm} - \langle x, z\rangle_{\pm}\right| \le \|y - z\|$
viii) $\langle sx, ry\rangle_{\pm} = r\langle x, y\rangle_{\pm}$ *for all* $r, s \ge 0$
  ix) $\langle x, \alpha x\rangle_{\pm} = \operatorname{Re}\alpha\|x\|$ *for all* $\alpha \in \mathbf{K}$
   x) $\langle x, y + \alpha x\rangle_{\pm} = \langle x, y\rangle_{\pm} + \operatorname{Re}\alpha\|x\|$ *for all* $\alpha \in \mathbf{K}$
  xi) *For every* $\varepsilon > 0$ *there exists* $h(\varepsilon) > 0$ *such that*

$$\langle x_1 + x_2, y\rangle_- \le \langle x_1, y\rangle_- + \|x_2\| h(\varepsilon) + \varepsilon.$$

$\square$

For each $x \in X$ we associate the sets

$$G(x) = \{g \in X^*\ ;\ \|g\| = 1,\ g(x) = \|x\|\}$$

and

$$F(x) = \left\{f \in X^*\ ;\ f(x) = \|x\|^2 = \|f\|^2\right\}.$$

The operator $F : X \mapsto \mathcal{P}(X^*)$ is called the *duality mapping* on X.

It is immediate that $F(x) = \|x\|G(x) = \{\|x\| \cdot f\ ;\ f \in G(x)\}$. The fact that $G(x)$ and $F(x)$ are nonempty is a consequence of Hahn–Banach theorem. Also, it is straightforward to see that $G(x)$ is a convex set for all $x \ne 0$ while $F(x)$ is convex for all $x \in X$.

A linear normed space X is called *strictly convex* if its unit sphere contains no line segments, i.e. $\|(1 - t)x + ty\| < 1$ for all $t \in\ ]0, 1[$ and $x, y \in S(0, 1)$ with $x \ne y$.

Proposition 1.1. *If X is a Banach space and its dual X^* is strictly convex, then for all $x \neq 0$, $G(x)$ consists of a single element.*

Proof. Suppose, for contradiction, that there are $f_1, f_2 \in G(x)$, $f_1 \neq f_2$. By convexity, $(1 - t)f_1 + tf_2 \in G(x) \subset S(0,1)$. But $S(0,1)$ does not contain line segments. $\square$

Let us consider now the subdifferential of the norm at the point $z \in X$ defined by
$$\partial \|z\| = \{f \in X^* \, ; \, \|z + x\| \geq \|z\| + \operatorname{Re} f(x) \text{ for all } x \in X\}.$$
We have the following characterizations of $\partial \|z\|$:

Lemma 1.2.
$$\partial \|z\| = \{f \in X^* \, ; \, \langle z, x \rangle_- \leq \operatorname{Re} f(x) \leq \langle z, x \rangle_+ \text{ for all } x \in X\}$$
$$= \{f \in X^* \, ; \, \operatorname{Re} f(x) \leq \langle z, x \rangle_+ \text{ for all } x \in X\}.$$

Proof. If $f \in \partial \|z\|$, $x \in X$, and $h > 0$ then
$$- h^{-1} (\|z - hx\| - \|z\|) \leq -h^{-1} (\|z\| + \operatorname{Re} f(-hx) - \|z\|) = \operatorname{Re} f(x) =$$
$$= h^{-1} (\|z\| + \operatorname{Re} f(hx) - \|z\|) \leq h^{-1} (\|z + hx\| - \|z\|).$$
Consequently,
$$\langle z, x \rangle_- \leq \operatorname{Re} f(x) \leq \langle z, x \rangle_+.$$
Conversely, suppose $\langle z, x \rangle_- \leq \operatorname{Re} f(x) \leq \langle z, x \rangle_+$ for all $x \in X$. Taking $h = 1$ in $\langle z, x \rangle_+ \leq h^{-1} (\|z + hx\| - \|z\|)$, it follows that
$$\|z + x\| - \|z\| \geq \langle z, x \rangle_+ \geq \operatorname{Re} f(x) \text{ for all } x \in X,$$
and hence $f \in \partial \|z\|$. This establishes the first equality. If $\operatorname{Re} f(x) \leq \langle z, x \rangle_+$ for all $x \in X$, then by i) of Lemma 1.1,
$$\operatorname{Re} f(x) = -\operatorname{Re} f(-x) \geq -\langle z, -x \rangle_+ = \langle z, x \rangle_-$$
and the final set equality is seen to be valid. $\square$

Proposition 1.2. *For every $f \in X^*$ the following properties are equivalent:*
   i) $\operatorname{Re} f(x) \leq \|x\|$ *for all* $x \in X$
   ii) $|f(x)| \leq \|x\|$ *for all* $x \in X$.

Proof. The implication ii) $\Rightarrow$ i) is trivial.

If we put $f(x) = |f(x)|e^{i\theta}$ then $|f(x)| = f(x)e^{-i\theta} = f(e^{-i\theta}x)$ is a real positive number. Hence,
$$|f(x)| = \operatorname{Re} f (e^{-i\theta}x) \leq \|e^{-i\theta}x\| = \|x\|.$$
$\square$

Now we are able to prove the following important result:

Lemma 1.3. *For every $z \in X$, $\partial \|z\| = G(z)$.*

Proof. Suppose first that $f \in \partial \|z\|$. By Lemma 1.2 and by parts i), vi) and ix) of Lemma 1.1 it follows that

$$\|z\| = \langle z, z \rangle_- \leq \operatorname{Re} f(z) \leq \langle z, z \rangle_+ = \|z\|$$

and

$$\operatorname{Re} f(x) \leq \langle z, x \rangle_+ \leq \|x\| \text{ for all } x \in X.$$

Therefore $\operatorname{Re} f(z) = \|z\|$ and the preceding proposition gives $|f(x)| \leq \|x\|$ for all $x \in X$, i.e. $f \in G(z)$.

Now suppose $f \in G(z)$. If $x \in X$ then

$$\operatorname{Re} f(z + x) \leq \|f(z + x)\| \leq \|z + x\|$$

and

$$\|z + x\| - \|z\| \geq \operatorname{Re} f(z + x) - \|z\| = \operatorname{Re} f(z) \ ;$$

i.e. $f \in \partial \|z\|$. □

The relation between the set $G(z)$ and the numbers $\langle z, y \rangle_{\pm}$ is of fundamental importance here. We have

Theorem 1.1. *If $z, y \in X$ and λ is a real number such that $\langle z, y \rangle_- \leq \lambda \leq \langle z, y \rangle_+$, then there is $f \in G(z)$ such that $\operatorname{Re} f(y) = \lambda$. In particular,*

$$\{\operatorname{Re} f(y) \ ; \ f \in G(z)\} = [\langle z, y \rangle_-, \langle z, y \rangle_+].$$

Proof. Taking X as a real linear space, let us consider the linear subspace of X spanned by z and y, that is,

$$D = \{\alpha z + \beta y \ ; \ \alpha, \beta \in \mathbf{R}\} \, .$$

We assume $y \neq \alpha z$ for any $\alpha \in \mathbf{R}$, since otherwise the assertion is immediate. Let us define the linear function $g : D \mapsto \mathbf{R}$ by $g(\alpha z + \beta y) = \alpha \|z\| + \beta \lambda$. If $\beta \geq 0$ then

$$\beta \lambda \leq \beta \langle z, y \rangle_+ = \langle z, \beta y \rangle_+$$

by Lemma 1.1, and if $\beta < 0$ then

$$\beta \lambda \leq \beta \langle z, y \rangle_- = -\beta \langle z, -y \rangle_+ = \langle z, \beta y \rangle_+$$

also by Lemma 1.1. Consequently, $\beta\lambda \leq \langle z, \beta y\rangle_+$ for all $\beta \in \mathbf{R}$ and we obtain by Lemma 1.1 that

$$g(\alpha z + \beta y) = \alpha\|z\| + \beta\lambda \leq \alpha\|z\| + \langle z, \beta y\rangle_+ =$$
$$= \langle z, \alpha z + \beta y\rangle_+ \leq \|\alpha z + \beta y\|.$$

By the Hahn–Banach theorem there is a linear function $h : X \mapsto \mathbf{R}$ such that $h(x) \leq \|x\|$ for all $x \in X$ and $h(x) = g(x)$ for all $x \in D$. Clearly h is continuous because $|h(x)| \leq \|x\|$ (see Proposition 1.2). Let us define $f : X \mapsto \mathbf{C}$ by $f(x) = h(x) - ih(ix)$. It is easy to show that $f \in X^*$. Because $\operatorname{Re} f(x) = h(x) \leq \|x\|$ for all $x \in X$, Proposition 1.2 gives $|f(x)| \leq \|x\|$ for every $x \in X$. On the other hand, $\operatorname{Re} f(z) = h(z) = g(z) = \|z\|$. Therefore, $f(z) = \operatorname{Re} f(z) = \|z\|$ such that $\|f\| = 1$ and $f \in G(z)$. Moreover, $\operatorname{Re} f(y) = h(y) = g(y) = \lambda$, and the first result is proved.

Let us take $f \in G(z)$ i.e. by Lemma 1.3, $f \in \partial\|z\|$. Hence, Lemma 1.2 implies

$$\langle z, y\rangle_- \leq \operatorname{Re} f(y) \leq \langle z, y\rangle_+ \text{ for all } y \in X.$$

This fact, together with the first part of the theorem give the required set equality. $\square$

Lemma 1.4. *Suppose* $\omega \in \mathbf{R}$ *and* $x, y \in X$. *Then the following statements are equivalent:*

    i) *There is* $f \in F(x)$ *such that* $\operatorname{Re} f(y) \leq \omega\|x\|^2$
    ii) *There is* $g \in G(x)$ *such that* $\operatorname{Re} g(y) \leq \omega\|x\|$
    iii) $\langle x, y\rangle_- \leq \omega\|x\|$
    iv) $\langle x, y - \omega x\rangle_- \leq 0$
    v) $\|x\| \leq \|(1 + h\omega)x - hy\|$ *for all* $h > 0$
    vi) $\|x\|(1 - \lambda\omega) \leq \|x - \lambda y\|$ *for all* $\lambda > 0$.

Proof. The equivalence of i) and ii) as well as iii) and iv) is immediate (see Lemma 1.1). If iv) is true then, Theorem 1.1 assures the existence of $g \in G(x)$ such that $\operatorname{Re} g(y - \omega x) \leq 0$ and ii) is proved. In the same manner we can see that ii) implies iv).

To prove the equivalence of ii) and v) we shall take $\omega = 0$. (The case $\omega \neq 0$ will be then proved by substituting y with $y - \omega x$.) Suppose first that v) is valid, i.e. $\|x\| \leq \|x - hy\|$ for all $h > 0$. Let us consider $g_h \in G(x - hy)$ with which we obtain

$$\|x\| \leq \|x - hy\| = g_h(x - hy) = \operatorname{Re} g_h(x) - h\operatorname{Re} g_h(y) \leq$$
$$\leq \|x\| - h\operatorname{Re} g_h(y)$$

from where

$$\operatorname{Re} g_h(y) \leq 0 \tag{1.1}$$

and

$$\|x\| \le \operatorname{Re} g_h(x) + h\|y\|.$$

Hence,

$$\|x\| \le \lim_{h \to 0} \inf \operatorname{Re} g_h(x). \tag{1.2}$$

Taking into account the weak*–compactness of $\overline{B(0,1)}$ in X^* (Alaoglu theorem, see Dunford and Schwartz [1958, p. 424]), we see that, on a subsequence, $g_h \xrightarrow{w^*} g$, where $\|g\| \le 1$. From (1.2) we derive $\|x\| \le \operatorname{Re} g(x) \le |g(x)| \le \|x\|$ such that $\operatorname{Re} g(x) = \|x\|$. Moreover $g(x) = \|x\|$ which implies $\|g\| = 1$ i.e. $g \in G(x)$. Finally from (1.1) we obtain $\operatorname{Re} g(y) \le 0$.

Conversely, if ii) is valid, i.e. $\operatorname{Re} g(y) \le 0$, then

$$\|x\| = g(x) = \operatorname{Re} g(x - hy) + \operatorname{Re} g(hy) \le \operatorname{Re} g(x - hy) \le |g(x - hy)| \le$$
$$\le \|x - hy\| \text{ for all } h > 0,$$

such that v) holds.

Now, if vi) is true we can write

$$\frac{\|x - \lambda y\| - \|x\|}{-\lambda} \le \frac{-\lambda \omega \|x\|}{-\lambda} = \omega \|x\|$$

and letting $\lambda \to 0^+$ one obtains iii). Finally, if ii) is valid we have

$$\|x\| = \operatorname{Re} g(x) = \operatorname{Re} g(x - \lambda y) + \operatorname{Re}(\lambda y) \le \|x - \lambda y\| + \lambda \omega \|x\|$$

which is just vi).

$\square$

Lemma 1.5. *If $\omega \in \mathbf{R}$ and $x, y \in X$, then the following assertions are equivalent:*

    i) *For all $f \in F(x)$, $\operatorname{Re} f(y) \le \omega \|x\|^2$*
    ii) *For all $g \in G(x)$, $\operatorname{Re} g(y) \le \omega \|x\|$*
    iii) *$\langle x, y \rangle_+ \le \omega \|x\|$*
    iv) *$\langle x, y - \omega x \rangle_+ \le 0$.*

Proof. The equivalence of i) and ii) is clear. Also, the equivalence of iii) and iv) is derived from Lemma 1.1 x).

To prove that ii) is equivalent with iv), means to prove that $\operatorname{Re} g(y - \omega x) \le 0$ for all $g \in G(x)$ is equivalent with iv). But this follows directly from Theorem 1.1.

$\square$

Let us consider now the following problem: when does $\langle x, y \rangle_- = \langle x, y \rangle_+$ for all $x, y \in X$ with $x \ne 0$? Proposition 1.1 combined with Theorem 1.1 shows that this fact holds if X^* is strictly convex.

For instance, if X is a Hilbert space with the inner product $\langle \, , \, \rangle$, then

$$F(x) = \{ f \in X^* \; ; \; f(y) = \langle x, y \rangle \text{ for all } y \in X \}$$

and Theorem 1.1 gives

$$\|x\| \langle x, y \rangle_- = \|x\| \langle x, y \rangle_+ = \mathrm{Re} \langle x, y \rangle \text{ for all } x, y \in X.$$

If we consider now a real Banach space with a strictly convex dual, by Theorem 1.1 we have a unique $f \in G(x)$ such that $\langle x, y \rangle_- = \langle x, y \rangle_+ = f(y)$ for all $x, y \in X$ and $x \neq 0$. But this means exactly the Gâteaux differentiability of the norm at the point $x \neq 0$. By Lemma 1.3, the Gâteaux derivative f of the norm is just its subdifferential. Also, we observe that the norm is never Gâteaux differentiable at $x = 0$ because

$$\langle 0, y \rangle_- \neq \langle 0, y \rangle_+ \text{ for all } y \in X \, , \, y \neq 0.$$

Now we shall compute the duality functionals for some spaces appearing in our applications.

To begin with, if Y is a compact space, let us consider the Banach space

$$C(Y; \mathbf{R}^n) = \{ f : Y \mapsto \mathbf{R}^n \text{ with continuous } f_i \text{ components} \}$$

with the usual "supremum" norm

$$\|f\| = \max_{1 \leq i \leq n} \max_{x \in Y} |f_i(x)|.$$

The following lemma is an extension of a theorem of Sato [1968] with a simpler proof (see Marinov, Neittaanmäki [1988]).

Lemma 1.6. For $f, g \in C(Y; \mathbf{R}^n)$, $f \neq 0$, it holds

$$\langle f, g \rangle_+ = \sup_{(p; x) \in M(f)} g_p(x) \, \mathrm{sgn} \, f_p(x)$$

where

$$M(f) = \left\{ (p, x) \; ; \; p \in \overline{1, n}, \, , x \in Y, \, |f_p(x)| = \|f\| \right\}.$$

Proof. Let $\{\varepsilon_k\}_k$ with $\varepsilon_k \to 0^+$ for $k \to \infty$. For every pair (k, i) of indices we choose $x_{k,i} \in Y$ such that

$$\max_{x \in Y} |f_i(x) + \varepsilon_k g_i(x)| = |f_i(x_{k,i}) + \varepsilon_k g_i(x_{k,i})|.$$

Consequently,

$$\|f + \varepsilon_k g\| = \max_{1 \le i \le n} |f_i(x_{k,i}) + \varepsilon_k g_i(x_{k,i})| =$$
$$= |f_p(x_{k,p}) + \varepsilon_k g_p(x_{k,p})|. \tag{1.3}$$

We choose from $\{x_{k,p}\}_k$ a convergent subsequence (keeping the same notation) $\{x_{k,p}\}_k$, such that $x_{k,p} \to x_p$, for $k \to \infty$. By (1.3), $(p, x_p) \in M(f)$. On the other hand, it is easy to see that there exists an index N such that for $k > N$ we have sgn $[f_p(x_{k,p}) + \varepsilon_k g_p(x_{k,p})] = $ sgn $f_p(x_{k,p}) = $ sgn $f_p(x_p)$. Hence, taking also into account the relation (1.3) we obtain for $k > N$:

$$\frac{\|f + \varepsilon_k g\| - \|f\|}{\varepsilon_k} \le \frac{|f_p(x_{k,p}) + \varepsilon_k g_p(x_{k,p})| - |f_p(x_{k,p})|}{\varepsilon_k} =$$
$$= g_p(x_{k,p}) \text{ sgn } f_p(x_p).$$

From here we find

$$\langle f, g \rangle_+ \le g_p(x_p) \text{ sgn } f_p(x_p) \le \sup_{(p;x) \in M(f)} g_p(x) \text{ sgn } f_p(x).$$

In order to prove the converse inequality, we take $(p, x_p) \in M(f)$ and we observe that

$$\frac{\|f + \varepsilon g\| - \|f\|}{\varepsilon} \ge |f_p(x_p)| \frac{\left| 1 + \varepsilon \dfrac{g_p(x_p)}{f_p(x_p)} \right| - 1}{\varepsilon}.$$

Due to the fact that the right hand side tends to $|f_p(x_p)| \dfrac{g_p(x_p)}{f_p(x_p)}$ when ε tends to 0^+, the inequality follows. □

The following two lemmas are slight extensions for complex valued functions of the results of Sato [1968].

Let Y be a normed space on which a σ-algebra and a measure m are defined and let L^p be the set of measurable complex functions h defined on Y for which

$$\|h\|^p = \int_Y |h|^p \, dm < \infty.$$

Lemma 1.7. For $f, g \in L^1$,

$$\langle f, g \rangle_+ = \int_{Y_0} |g| \, dm + \int_{Y-Y_0} |f| \operatorname{Re} \frac{g}{f} \, dm$$

where $Y_0 = \{x \in Y \; ; \; f(x) = 0\}$.

Proof. We have

$$
\begin{aligned}
\langle f, g \rangle_+ &= \lim_{\varepsilon \to 0+} \frac{\int_Y |f + \varepsilon g|\, dm - \int_Y |f|\, dm}{\varepsilon} \\
&= \lim_{\varepsilon \to 0+} \frac{\varepsilon \int_{Y_0} |g|\, dm + \int_{Y-Y_0} |f + \varepsilon g|\, dm - \int_{Y-Y_0} |f|\, dm}{\varepsilon} \\
&= \int_{Y_0} |g|\, dm + \lim_{\varepsilon \to 0+} \int_{Y-Y_0} |f| \frac{\left| 1 + \varepsilon \frac{g}{f} \right| - 1}{\varepsilon}\, dm.
\end{aligned}
$$

But,

$$
|f(x)| \frac{\left| 1 + \varepsilon \frac{g(x)}{f(x)} \right| - 1}{\varepsilon} \leq |g(x)|
$$

and

$$
\lim_{\varepsilon \to 0+} \frac{\left| 1 + \varepsilon \frac{g(x)}{f(x)} \right| - 1}{\varepsilon} = \operatorname{Re} \frac{g(x)}{f(x)}, \quad \text{for every } x \in Y - Y_0.
$$

Then, by Lebesgue's Dominated Convergence theorem we have the result.

Lemma 1.8. For any p, $1 < p < \infty$, and $f, g \in L^p$, $f \neq 0$

$$
\langle f, g \rangle_+ = \frac{1}{\|f\|^{p-1}} \int_{Y-Y_0} |f|^p \operatorname{Re} \frac{g}{f}\, dm,
$$

where $Y_0 = \{x \in Y \, ; \, f(x) = 0\}$.

Proof. The right side of the above equality is linear in g and, by Hölder's inequality, is majorized by $\|g\|$ in the absolute value. Since $\langle f, g \rangle_+$ is also continuous in g (see Lemma 1.1 vii)), it suffices to prove the result for dense g's. Hence we suppose that there are positive constants c and δ such that $|g(x)| \leq c$ on Y and $g(x) = 0$ on the set Y_1 of points x where $0 < |f(x)| < \delta$. Let $Y_2 = \{x \in Y \, ; \, |f(x)| \geq \delta\}$. We have, for all $\varepsilon > 0$

$$
\int_Y |f + \varepsilon g|^p\, dm = \varepsilon^p \int_{Y_0} |g|^p\, dm + \int_{Y_1} |f|^p\, dm + \int_{Y_2} |f + \varepsilon g|^p\, dm.
$$

But

$$
\int_{Y_2} |f + \varepsilon g|^p\, dm = \int_{Y_2} |f|^p \left(1 + 2\varepsilon F_r + \varepsilon^2 |F|^2 \right)^{\frac{p}{2}}\, dm,
$$

where $F = g/f$ and $F_r = \operatorname{Re} F$. Since on Y_2 we have $F_r \leq |F| \leq c/\delta$ and

$$
\left| 2\varepsilon F_r + \varepsilon^2 |F|^2 \right| \leq 2\varepsilon \frac{c}{\delta} + \varepsilon^2 \frac{c^2}{\delta^2}
$$

it holds

$$\int_{Y_2} |f + \varepsilon g|^p \, dm = \int_{Y_2} |f|^p \left[1 + \frac{p}{2} \left(2\varepsilon F_r + \varepsilon^2 |F|^2 \right) + o(\varepsilon) \right] \, dm =$$

$$= \int_{Y_2} |f|^p \left(1 + \varepsilon p F_r + o(\varepsilon) \right) \, dm.$$

It follows that

$$\varepsilon^{-1} \left(\|f + \varepsilon g\| - \|f\| \right) =$$

$$= \varepsilon^{-1} \left[\int_Y |f|^p \, dm + \varepsilon p \int_{Y_1 \cup Y_2} F_r |f|^p \, dm + o(\varepsilon) \right]^{\frac{1}{p}} - \varepsilon^{-1} \left(\int_Y |f|^p \, dm \right)^{\frac{1}{p}}.$$

But $(\int_{Y_1 \cup Y_2} F_r |f|^p \, dm) / \|f\|^p \le \frac{c}{\delta}$ such that

$$\varepsilon^{-1} \left(\|f + \varepsilon g\| - \|f\| \right) = \|f\| \frac{1 + \varepsilon \left(\int_{Y_1 \cup Y_2} F_r |f|^p \, dm \right) / \|f\|^p + o(\varepsilon) - 1}{\varepsilon}$$

from where we obtain the desired result.   □

1.2. Dissipative operators

Let us consider $A : \mathcal{D}(A) \mapsto X$, $\mathcal{D}(A) \subset X$, with its "range", $\mathcal{R}(A) = \{Ax \; ; \; x \in \mathcal{D}(A)\}$.

If $\omega \in \mathbf{R}$, A is said to be ω–*dissipative* if for all $x, y \in \mathcal{D}(A)$, any of the following equivalent properties (see Lemma 1.4) are valid:

D1. There is $f \in F(x - y)$ such that

$$\operatorname{Re} f \left(Ax - Ay \right) \le \omega \|x - y\|^2.$$

D2. There is $g \in G(x - y)$ such that

$$\operatorname{Re} g \left(Ax - Ay \right) \le \omega \|x - y\|.$$

D3. $\langle x - y, Ax - Ay \rangle_- \le \omega \|x - y\|$.

D4. $\langle x - y, Ax - Ay - \omega(x - y) \rangle_- \le 0$.

D5. $\|x - y\| \le \|(1 + h\omega)(x - y) - h(Ax - Ay)\|$ for all $h > 0$.

D6. $\|x - y\|(1 - \lambda\omega) \le \|x - y - \lambda(Ax - Ay)\|$ for all $\lambda > 0$.

The operator A will be called *dissipative* if it is 0–dissipative, and *strongly dissipative* if it is ω–dissipative, where $\omega < 0$.

A is called *accretive* (*monotone*, if X is a Hilbert space) if $-A$ is dissipative.

Let us observe that, by the definition **D4**, ω–dissipativity of A means exactly the dissipativity of $A - \omega\mathcal{I}$, where $\mathcal{I}$ is the identity operator on X.

A is said to be *totally ω-dissipative* if for all $x, y \in \mathcal{D}(A)$, any of the following equivalent properties (see Lemma 1.5) are valid:

D7. For all $f \in F(x - y)$, $\mathrm{Re}\, f(Ax - Ay) \leq \omega\|x - y\|^2$.

D8. For all $g \in G(x - y)$, $\mathrm{Re}\, g(Ax - Ay) \leq \omega\|x - y\|$.

D9. $\langle x - y, Ax - Ay\rangle_+ \leq \omega\|x - y\|$.

D10. $\langle x - y, Ax - Ay - \omega(x - y)\rangle_+ \leq 0$.

If we call A *totally dissipative* when it is totally 0–dissipative, then by **D10**, total ω–dissipativity of A is equivalent with total dissipativity of $A - \omega\mathcal{I}$. Also, by Lemma 1.1 vi) it is clear that total ω–dissipativity implies the ω–dissipativity of A.

A is said to be *"m"–dissipative* if A is dissipative and there is $\alpha > 0$ such that $\mathcal{R}(\mathcal{I} - \alpha A) = X$. Similarly, we introduce the notions of *"m"–total dissipativity* and *"m"–accretivity*.

Proposition 1.3. *If A is dissipative then, for all $\alpha > 0$, $(\mathcal{I} - \alpha A)^{-1}$ is well defined and it is a contraction on $\mathcal{R}(\mathcal{I} - \alpha A)$.*

Proof. If we denote $B = \mathcal{I} - \alpha A$, then the dissipativity of A implies (see **D5**)

$$\|x - y\| \leq \|Bx - By\| \text{ for all } x, y \in \mathcal{D}(A) \tag{1.4}$$

that is the injectivity of B and the contractivity of B^{-1}. $\square$

Lemma 1.9. *A is "m"–dissipative if and only if A is dissipative and $\mathcal{R}(\mathcal{I} - \alpha A) = X$ for all $\alpha > 0$.*

Proof. Actually we have to prove that $\mathcal{R}(\mathcal{I} - A) = X$ implies $\mathcal{R}(\mathcal{I} - \alpha A) = X$ for all $\alpha > 0$ with condition

$$\|(\mathcal{I} - A)^{-1}x - (\mathcal{I} - A)^{-1}y\| \leq \|x - y\| \text{ for all } x, y \in X, \tag{1.5}$$

(see (1.4)).

Take an arbitrary $v \in X$. We must show that there is $u \in \mathcal{D}(A)$ such that $v = u - \alpha Au$, or equivalently

$$u = (\mathcal{I} - A)^{-1}\left[\alpha^{-1}v + \alpha^{-1}(\alpha - 1)u\right].$$

Define $T : X \mapsto \mathcal{D}(A)$, by

$$Tu = (\mathcal{I} - A)^{-1} \left[\alpha^{-1} v + \alpha^{-1} (\alpha - 1) u \right].$$

Then (1.5) implies that T is Lipschitz continuous with Lipschitz constant $|\alpha^{-1} - 1|$. Therefore, for each $\alpha > 1/2$, $|\alpha^{-1} - 1| < 1$ so T has a fixed point. Thus we have proved the result for each $\alpha > 1/2$. Since αA is also dissipative (for any $\alpha > 0$), we can apply the previous result to $\alpha A, \alpha^2 A, \ldots, \alpha^n A, \ldots$

In other words, we have proved that $\mathcal{R}(\mathcal{I} - A) = X$ implies

$$\mathcal{R}(\mathcal{I} - \alpha^n A) = X, \text{ for all } n \in \mathbb{N} \text{ and } \alpha > \frac{1}{2},$$

from which we easily derive (see Oharu [1966, p. 1150]) the desired result.   □

The dissipative operator A is said to be *maximal dissipative* if whenever $x_0, y_0 \in X$ satisfies

$$\langle x - x_0, Ax - y_0 \rangle_- \leq 0 \text{ for all } x \in \mathcal{D}(A),$$

we obtain $x_0 \in \mathcal{D}(A)$ and $y_0 = Ax_0$.

Proposition 1.4. *If A is "m"–dissipative then A is maximal dissipative.*

Proof. Assume by contradiction that there exists $x_0 \notin \mathcal{D}(A)$ and $y_0 \in X$ such that

$$\langle x - x_0, Ax - y_0 \rangle_- \leq 0 \text{ for all } x \in \mathcal{D}(A). \tag{1.6}$$

Since $\mathcal{R}(\alpha \mathcal{I} - A) = X$, there exists $x_1 \in \mathcal{D}(A)$ such that

$$\alpha x_1 - A x_1 = \alpha x_0 - y_0. \tag{1.7}$$

Now we take in (1.6) $x = x_1$ and following (1.7) we obtain

$$\langle x_1 - x_0, \alpha(x_1 - x_0) \rangle_- \leq 0.$$

By Lemma 1.1 ix) we find $x_0 = x_1 \in \mathcal{D}(A)$ and from (1.7), $y_0 = Ax_1 = Ax_0$, that is a contradiction.   □

The above proposition justifies the usage of notion *hyper–maximal dissipative operator* instead of "m"–dissipative operator.

1.3. Semigroups of linear operators

If X is a Banach space over $\mathbf{K}$ ($\mathbf{R}$ or $\mathbf{C}$) and $\mathcal{D}(A)$ is a linear subspace of X, let us consider $A : \mathcal{D}(A) \mapsto X$ a linear (unbounded) operator. The *resolvent set* of A is

$$\rho(A) = \{\lambda \in \mathbf{C} \; ; \; \mathcal{R}(\lambda \mathcal{I} - A) = X \text{ and}$$

$$\text{there exists the bounded operator } (\lambda \mathcal{I} - A)^{-1}\}.$$

A C_0-*semigroup* S on X is a family of linear bounded operators $S = \{S(t) \; ; \; t \in [0, \infty[\}$ satisfying

    i) $S(t)S(s) = S(t+s)$ for each $t, s \in [0, \infty[$
   ii) $S(0) = \mathcal{I}$
  iii) $S(\cdot)x : [0, \infty[\mapsto X$ is continuous for each $x \in X$.

A C_0-*contraction semigroup* S on X is a C_0-semigroup such that for each $t \in [0, \infty[, \|S(t)\| \leq 1$.

Let S be a C_0-semigroup on X. The *generator* A of S is defined by the formula

$$Ax = \lim_{t \to 0^+} \frac{S(t)x - x}{t} = \frac{d^+}{dt} S(t)x \bigg|_{t=0}$$

and the domain $\mathcal{D}(A)$ of A is the set of all $x \in X$ for which the above limit exists.

Lemma 1.10. *Let A_1 and A_2 generate contraction semigroups S_1 and S_2, respectively, such that $S_1(t)S_2(s) = S_2(s)S_1(t)$ for all $s, t \in [0, \infty[$. Then, for each $x \in \mathcal{D}(A_1) \cap \mathcal{D}(A_2)$*

$$\|S_1(t)x - S_2(t)x\| \leq t\|A_1 x - A_2 x\|.$$

Proof. As

$$S_1(t)x - S_2(t)x = \int_0^1 \frac{d}{ds} \left(S_1(ts)S_2(t(1-s))x\right) ds$$

$$= \int_0^1 S_1(ts)S_2(t(1-s))t(A_1 x - A_2 x) ds,$$

the result follows. $\square$

Lemma 1.11. *Let S be a closed and densely defined linear operator on X, and let $\mu \in \rho(S)$. Then $\lambda \in \rho(S)$ if and only if $\mathcal{I} - (\mu - \lambda)(\mu \mathcal{I} - S)^{-1}$ has a bounded inverse; in this case*

$$(\lambda \mathcal{I} - S)^{-1} = (\mu \mathcal{I} - S)^{-1} \left[\mathcal{I} - (\mu - \lambda)(\mu \mathcal{I} - S)^{-1}\right]^{-1}.$$

Proof. Suppose $T = \mathcal{I} - (\mu - \lambda)(\mu\mathcal{I} - S)^{-1}$ has a bounded inverse. Then

$$(\lambda\mathcal{I} - S)(\mu\mathcal{I} - S)^{-1}T^{-1} = [(\lambda - \mu)\mathcal{I} + (\mu\mathcal{I} - S)](\mu\mathcal{I} - S)^{-1}T^{-1} =$$
$$= [(\lambda - \mu)(\mu\mathcal{I} - S)^{-1} + \mathcal{I}]T^{-1} = \mathcal{I}$$

and similarly,

$$(\mu\mathcal{I} - S)^{-1}T^{-1}(\lambda\mathcal{I} - S) = \mathcal{I}|_{\mathcal{D}(S)}.$$

Thus $\lambda \in \rho(S)$ and the desired equality holds.

The proof of the converse is equally easy.

Now, we are ready to prove the main result about linear semigroups.

Theorem 1.2. *(Hille–Yosida)* A *is the generator of a* C_0*–contraction semigroup if and only if* A *is closed, densely defined,* $]0, \infty[\subset \rho(A)$ *and* $\|(\lambda\mathcal{I} - A)^{-1}\| \le 1/\lambda$ *for all* $\lambda > 0$.

Proof. **A.** Necessity.

For each $x \in \mathcal{D}(A)$,

$$\frac{d^+}{dt}S(t)x = \lim_{h \to 0+} \frac{S(t + h) - S(t)}{h}x = \lim_{h \to 0+} S(t)\frac{S(h) - \mathcal{I}}{h}x = S(t)Ax =$$
$$= \lim_{h \to 0+} \frac{S(h) - \mathcal{I}}{h}S(t)x = AS(t)x.$$

Thus $S(t)(\mathcal{D}(A)) \subset \mathcal{D}(A)$ and

$$\frac{d^+}{dt}S(t)x = AS(t)x = S(t)Ax \ , \ x \in \mathcal{D}(A). \tag{1.8}$$

Also, if $t > 0$ we can similarly derive for $x \in \mathcal{D}(A)$

$$\frac{d^-}{dt}S(t)x = AS(t)x = S(t)Ax. \tag{1.9}$$

Let us remark in passing that, (1.8) and (1.9) show that $u(\cdot) = S(\cdot)x$ solves the initial value problem

$$\begin{cases} \dfrac{du}{dt} = Au(t) \ , \ t \ge 0 \ , \\ u(0) = x \end{cases} \tag{1.10}$$

when A is the generator of S and $x \in \mathcal{D}(A)$.

Thus for each $x \in \mathcal{D}(A)$, $S(\cdot)x \in C^1\,(]0,\infty[; X)$ and

$$S(t)x - x = \int_0^t \frac{d}{ds} S(s)x\,ds = \int_0^t AS(s)x\,ds = \int_0^t S(s)Ax\,ds. \qquad (1.11)$$

Let $x \in X$ and set $x_t = \int_0^t S(s)x\,ds$. Clearly $\lim_{t \to 0+} t^{-1}x_t = x$ and

$$h^{-1}\,(S(h) - \mathcal{I})\,x_t = h^{-1} \int_t^{t+h} S(s)x\,ds - h^{-1} \int_0^h S(s)x\,ds$$
$$\to S(t)x - x\ (= Ax_t)\ \text{as}\ h \to 0^+.$$

Thus $x_t \in \mathcal{D}(A)$ and so $\overline{\mathcal{D}(A)} = X$. Moreover, we have shown

$$S(t)x - x = A \int_0^t S(s)x\,ds \ \text{for all}\ x \in X. \qquad (1.12)$$

Let $x_n \in \mathcal{D}(A)$, such that $x_n \to x$, $Ax_n \to f$. Then

$$t^{-1}(S(t) - \mathcal{I})x_n = t^{-1} \int_0^t S(s)Ax_n\,ds$$

by (1.11). When $n \to \infty$ we obtain

$$t^{-1}(S(t) - \mathcal{I})x = t^{-1} \int_0^t S(s)f\,ds \to f$$

as $t \to 0^+$. Thus $x \in \mathcal{D}(A)$ and $Ax = f$ i.e. A is closed.

For each $\lambda > 0$, $\{e^{-\lambda t}S(t)\ ;\ t \in]0,\infty[\}$ is a C_0–contraction semigroup with generator $A - \lambda \mathcal{I}$. Applying (1.12) and (1.11) to this semigroup gives

$$e^{-\lambda t}S(t)x - x = A - (\lambda \mathcal{I}) \int_0^t e^{-\lambda s}S(s)x\,ds\ , \quad x \in X$$
$$e^{-\lambda t}S(t)x - x = \int_0^t e^{-\lambda s}S(s)(A - \lambda \mathcal{I})x\,ds\ , \quad x \in \mathcal{D}(A).$$

Now let $t \to \infty$; the closedness of A and the Dominated Convergence theorem imply $\int_0^\infty e^{-\lambda s}S(s)x\,ds \in \mathcal{D}(A)$ and

$$x = (\lambda \mathcal{I} - A) \int_0^\infty e^{-\lambda s}S(s)x\,ds\ , \quad x \in X$$
$$x = \int_0^\infty e^{-\lambda s}S(s)(\lambda \mathcal{I} - A)x\,ds\ , \quad x \in \mathcal{D}(A).$$

Thus $\lambda \in \rho(A)$ and

$$(\lambda \mathcal{I} - A)^{-1} y = \int_0^\infty e^{-\lambda s} S(s) y \, ds \ , \ y \in X \ , \ \lambda > 0. \tag{1.13}$$

Moreover,

$$\left\| (\lambda \mathcal{I} - A)^{-1} y \right\| \le \int_0^\infty e^{-\lambda s} \| S(s) \| \cdot \| y \| \, ds \le \| y \| / \lambda.$$

This completes the proof of the necessity.

B. Sufficiency.

For $\lambda > 0$, set

$$A_\lambda = \lambda A (\lambda \mathcal{I} - A)^{-1} = \lambda^2 (\lambda \mathcal{I} - A)^{-1} - \lambda \mathcal{I}.$$

This bounded operator is called *Yosida-approximation* of A. This is because,

$$\lim_{\lambda \to \infty} A_\lambda x = A x \text{ for all } x \in \mathcal{D}(A).$$

We have $\lambda (\lambda \mathcal{I} - A)^{-1} x - (\lambda \mathcal{I} - A)^{-1} A x = x$, $(\lambda \mathcal{I} - A)^{-1} A x \to 0$ and so $\lambda (\lambda \mathcal{I} - A)^{-1} x \to x$ as $\lambda \to \infty$, for all $x \in \mathcal{D}(A)$ and hence for all $x \in \overline{\mathcal{D}(A)} = X$. Thus if $x \in \mathcal{D}(A)$, $A_\lambda x = \lambda (\lambda \mathcal{I} - A)^{-1} A x \to A x$ as $\lambda \to \infty$.

Let us define the family of operators

$$\left\{ e^{t A_\lambda} = \lim_{n \to \infty} (\mathcal{I} - t A_\lambda / n)^{-n} ; \ t \in [0, \infty[\right\}$$

which is a C_0–contraction semigroup generated by A_λ.

By applying Lemma 1.10 with $A_1 = A_\lambda$, $A_2 = A_\mu$, we have

$$\left\| e^{t A_\lambda} x - e^{t A_\mu} x \right\| \le t \| A_\lambda x - A_\mu x \| \to 0$$

as $\lambda, \mu \to \infty$, for each $x \in \mathcal{D}(A)$ (t fixed). Define

$$S(t) x = \lim_{\lambda \to \infty} e^{t A_\lambda} x \ , \ x \in \mathcal{D}(A).$$

Clearly $\| S(t) \| \le 1$, and the above equation holds for all $x \in X$. Moreover $S(t) S(s) = S(t + s)$, $S(0) = \mathcal{I}$. Next, for $x \in \mathcal{D}(A)$

$$\begin{aligned} S(t) x - x &= \lim_{\lambda \to \infty} e^{t A_\lambda} x - x = \\ &= \lim_{\lambda \to \infty} \int_0^t e^{s A_\lambda} A_\lambda x \, ds = \int_0^t S(s) A x \, ds \end{aligned} \tag{1.14}$$

by the bounded convergence theorem. Thus $S(\cdot) x$ is continuous on $[0, \infty[$ for each $x \in \mathcal{D}(A)$ and hence for each $x \in X$. Thus S is a C_0–contraction semigroup. Let B denote its generator. Then (1.14) implies $B \supset A$ i.e. $\mathcal{D}(B) \supset \mathcal{D}(A)$ and $B|_{\mathcal{D}(A)} = A$. By the necessity part of the theorem, $1 \in \rho(B)$; also $1 \in \rho(A)$. Hence $(\mathcal{I} - B)^{-1} = (\mathcal{I} - A)^{-1}$ since $(\mathcal{I} - B)^{-1} \supset (\mathcal{I} - A)^{-1}$ and both are bounded operators. If follows that $B = A$. This completes the proof. □

Now, we give an alternative formulation of the Hille–Yosida theorem due to Lumer and Phillips [1961].

Theorem 1.3. *(Lumer–Phillips)*

    a) *Suppose A generates a C_0–contraction semigroup. Then*

        i) $\overline{\mathcal{D}(A)} = X$

        ii) *A is totally dissipative*

        iii) $]0, \infty[\subset \rho(A)$.

    b) *Conversely, if A satisfies*

        i') $\overline{\mathcal{D}(A)} = X$

        ii') *A is dissipative*

        iii') $]0, \infty[\cap \rho(A) \neq \emptyset$

    then A generates a C_0–contraction semigroup on X.

Proof.

    a) By Theorem 1.2, i) and iii) hold. Let $x \in \mathcal{D}(A)$. If $g \in G(x)$,

$$\operatorname{Re} g\left(S(t)x - x\right) = \operatorname{Re} g\left(S(t)x\right) - \|x\| \leq$$
$$\leq |g\left(S(t)x\right)| - \|x\| \leq \|S(t)x\| - \|x\| \leq 0.$$

If we divide by t and let $t \to 0$, ii) follows.

    b) Let $\lambda > 0$, $0 \neq x \in \mathcal{D}(A)$. Then by definition D5 of dissipativity, $\lambda \|x\| \leq \|(\lambda \mathcal{I} - A)x\|$. Thus, $\lambda \mathcal{I} - A$ is injective and $\|(\lambda \mathcal{I} - A)^{-1}\| \leq 1/\lambda$ on $\mathcal{R}(\lambda \mathcal{I} - A)$. A is closed and hence $\mathcal{R}(\lambda \mathcal{I} - A)$ is closed. It remains to show that $\mathcal{R}(\lambda \mathcal{I} - A) = X$, i.e. $\lambda \in \rho(A)$ for each $\lambda > 0$. The result then follows as a consequence of the sufficiency part of Theorem 1.2. By assumption iii'), there is some $\mu \in]0, \infty[\cap \rho(A)$. Also, note that $\|\mu(\mu \mathcal{I} - A)^{-1}\| \leq 1$. If $|\alpha| < \mu$ then $\|\alpha(\mu \mathcal{I} - A)^{-1}\| \leq |\alpha|/\mu < 1$ so that $\mathcal{I} - \alpha(\mu \mathcal{I} - A)^{-1}$ is invertible and its inverse is also bounded. By Lemma 1.11, $\lambda \in \rho(A)$ if $|\lambda - \mu| < \mu$, i.e. if $0 < \lambda < 2\mu$. Applying Lemma 1.11 again (with $3\mu/2$ in place of μ) we get $]0, 3\mu[\subset \rho(A)$. Thus, the proof is ended by an induction argument. $\qquad\square$

    A C_0–semigroup S is said to be *differentiable* if $S(t)x \in \mathcal{D}(A)$, for all $x \in X$ and $t > 0$, A being the generator of S. This definition becomes clear in the following result:

Lemma 1.12. *If the C_0–semigroup S is differentiable, then*

    a) *$t \mapsto S(t)x$ is continuously differentiable from $]0, \infty[$ into X.*

    b) *$\dfrac{d}{dt} S(t)x = AS(t)x$ for all $t > 0$ and $x \in X$.*

Proof.

$$\frac{d^+}{dt} S(t)x = \lim_{h \to 0^+} h^{-1} \left[S(h)S(t)x - S(t)x\right] = AS(t)x$$

and

$$\frac{d^-}{dt}S(t)x = \lim_{h\to 0+} S(2^{-1}t - h)h^{-1}\left[S(2^{-1}t + h)x - S(2^{-1}t)x\right] =$$
$$= \lim_{h\to 0+} S(2^{-1}t - h)h^{-1}\left[S(h)S(2^{-1}t)x - S(2^{-1}t)x\right] =$$
$$= S(2^{-1}t)AS(2^{-1}t)x = AS(t)x.$$

Moreover, if $t \geq \delta > 0$ then $S(\delta)x \in \mathcal{D}(A)$, $AS(t)x = S(t - \delta)AS(\delta)x$ and it follows that the map $t \mapsto AS(t)x$ is continuous from $]0, \infty[$ into X.   □

A C_0–semigroup S generated by A is said to be *analytic* if S is differentiable and there is a number $N > 0$ such that

$$t\|AS(t)\| \leq N \text{ for all } t \in]0, 1].$$

This definition is equivalent to the existence of a number $\alpha \in]0, \frac{\pi}{2}[$ such that the semigroup $S = \{S(t) \;;\; t > 0\}$ has an analytic extension $\tilde{S} = \{\tilde{S}(t) \;;\; t \in \mathbb{C}, \mid \arg (t)\mid < \alpha\}$ (see Butzer and Berens [1967, Proposition 1.1.11]).

A useful criterion, in the case when X is a complex Banach space, for operator A to be a generator of an analytic semigroup, is the following one (see Fattorini [1983, Corollary 4.2.5]):

Lemma 1.13. *Let A be a densely defined, "m"-dissipative linear operator on X, such that for all $x \in \mathcal{D}(A)$ there exist $f \in F(x)$, $\delta \geq 0$ with*

$$\mathrm{Re}\, f(Ax) \leq -\delta\, |\, \mathrm{Im}\, f(Ax)|.$$

Then A generates an analytic C_0-contraction semigroup.   □

1.4. Linear differential equations on Banach spaces

Let us consider the Banach space X over the field $\mathbf{K}$ and $A : \mathcal{D}(A) \subset X \mapsto X$, a linear operator on X.

We consider in this section the existence and uniqueness of the solution for the Cauchy problem in X:

$$\begin{cases} \dfrac{du}{dt} = Au(t) + f(t) \ , \ t > 0 \\ u(0) = u_0 \ . \end{cases} \tag{1.15}$$

A function $u \in C^1(]0, \infty[; X)$ verifying $u(t) \in \mathcal{D}(A)$ for all $t > 0$ such that (1.15) holds, is said to be a *classical solution* of (1.15) or, simply, a *solution*.

A function $u \in C(]0, \infty[; X)$ is called a *strong solution* of (1.15) if it is absolutely continuous on every compact of $]0, \infty[$, $u(t) \in \mathcal{D}(A)$ and satisfies (1.15) a.e. on $]0, \infty[$.

Theorem 1.4. *Let $\mathcal{D}(A)$ be a subspace of X such that $\overline{\mathcal{D}(A)} = X$ and $A : \mathcal{D}(A) \mapsto X$ be a linear "m"-dissipative operator. Let also $f \in C^1([0,\infty[; X)$ and $u_0 \in \mathcal{D}(A)$. Then, the Cauchy problem (1.15) has a unique classical solution.*

Proof. Taking into account Proposition 1.3 we see that the "m"-dissipativity of A implies $]0,\infty[\subset \rho(A)$. Then by Theorem 1.3 b), A generates a C_0-contraction semigroup denoted by $S(t)$. If u is a solution, then

$$\frac{d}{ds}(S(t-s)u(s)) = S(t-s)f(s)$$

and we obtain

$$u(t) = S(t)u_0 + \int_0^t S(t-s)f(s)\, ds. \tag{1.16}$$

This suggests that we seek a solution of the form (1.16).

Let

$$v(t) = \int_0^t S(t-s)f(s)\, ds .$$

Clearly $u \in C^1([0,\infty[; X)$ iff $v \in C^1([0,\infty[; X)$, in which case $u'(t) = AS(t)u_0 + v'(t)$. This means that (1.15) holds iff

$$v'(t) = Av(t) + f(t). \tag{1.17}$$

Recall that for all $x \in X$, $\int_a^b S(t)x\, dt \in \mathcal{D}(A)$ and

$$A \int_a^b S(t)x\, dt = S(b)x - S(a)x .$$

Then

$$
\begin{aligned}
v(t) &= \int_0^t S(t-s)\left[f(0) + \int_0^s f'(r)\, dr\right]\, ds = \\
&= \int_0^t S(t-s)f(0)\, ds + \int_0^t \int_r^t S(t-s)f'(r)\, ds dr = \\
&= \int_0^t S(\sigma)f(0)\, d\sigma + \int_0^t \int_0^{t-r} S(\sigma)f'(r)\, d\sigma dr \in \mathcal{D}(A) .
\end{aligned}
\tag{1.18}
$$

Also $v \in C^1([0,\infty[; X)$ and

$$
\begin{aligned}
v'(t) &= \frac{d}{dt} \int_0^t S(\sigma)f(t-\sigma)\, d\sigma = \\
&= S(t)f(0) + \int_0^t S(\sigma)f'(t-\sigma)\, d\sigma = \\
&= S(t)f(0) + \int_0^t S(t-s)f'(s)\, ds.
\end{aligned}
\tag{1.19}
$$

Otherwise, by (1.18)

$$Av(t) = S(t)f(0) - f(0) + \int_0^t [-f'(r) + S(t - r)f'(r)] \, dr =$$

$$= S(t)f(0) - f(0) - f(t) + f(0) + \int_0^t S(t - r)f'(r) \, dr$$

which combined with (1.19) gives (1.17). The existence is proved.

For the uniqueness, let u and v be two solutions of (1.15). Then $w = u - v$ satisfies

$$\begin{cases} \dfrac{dw}{dt} = Aw(t) \, , \; t \geq 0; \\ w(0) = 0 \, . \end{cases}$$

If $S(t)$ is the semigroup generated by A, we have

$$\frac{d}{ds} S(t - s)w(s) = S(t - s)Aw(s) - S(t - s)Aw(s) = 0$$

whence $S(t-s)w(s)$ is independent of s; setting $s = 0$, $s = t$ yields $w(t) = w(0) = 0$ for all t. This completes the proof. $\qquad\square$

It is not difficult to show, essentially with the same proof as above, that we have:

Theorem 1.5. (Pazy [1983, Theorem 2.9]) *Let A be the generator of a C_0-semigroup. If f is differentiable a.e. on $[0,T]$ and $f' \in L^1(0,T \; ; \; X)$ then for every $u_0 \in \mathcal{D}(A)$ the initial value problem (1.15) has a unique strong solution on $[0,T]$.* $\qquad\square$

Also, starting with the "mild" solution (1.16) it is easy to prove the so–called "smoothing effect on initial data". This means that the solution may be of the classical type even in the case $u_0 \notin \mathcal{D}(A)$. Precisely, we have the following result (Martin [1976, Proposition 4.2, Ch. 7]):

Theorem 1.6. *Suppose A is the generator of a differentiable semigroup and $f \in C^1 \, ([0, \infty[\; ; \; X)$. Then, for each $u_0 \in X$, the problem (1.15) has a unique classical solution.* $\qquad\square$

Moreover, this property holds for f being a Hölder continuous function. For $\nu \in]0,1]$, $C^\nu(0,T \; ; \; X)$ will denote the space of functions for which there exists $M > 0$ such that $\|f(t_1) - f(t_2)\| \leq M|t_1 - t_2|^\nu$ for all $t_1, t_2 \in]0,T[$.

Theorem 1.7. *Let A be the generator of an analytic semigroup and $f \in C^{\nu}(0, T; X)$ for each $T > 0$. Then, for every $u_0 \in X$, the problem (1.15) has a unique classical solution.*

Proof. Let $S(t)$ be the semigroup generated by A. Since the map $t \mapsto S(t)u_0$ is continuously differentiable on $[0, \infty[$, it suffices to show that

$$v(t) = \int_0^t S(t - s)f(s)\, ds \in \mathcal{D}(A)$$

and $t \mapsto Av(t)$ is continuous on $[0, \infty[$.

So let $T > 0$, $M > 0$ and $\nu \in]0, 1]$ be such that $\|f(t) - f(s)\| \le M|t - s|^{\nu}$ for $t, s \in [0, T]$. Set

$$v_1(t) = \int_0^t S(t - s)[f(s) - f(t)]\, ds$$

and

$$v_2(t) = \int_0^t S(t - s)f(t)\, ds \quad \text{for all } t \ge 0$$

and note $v = v_1 + v_2$. Moreover, it is easy to see that $v_2(t) \in \mathcal{D}(A)$ and $Av_2(t) = S(t)f(t) - f(t)$ is continuous on $[0, \infty[$. Since S is analytic, let M be large enough so that $\|S(t)\| \le M$ and $\|AS(t)\| \le t^{-1}M$ for all $t \in]0, T]$. For each positive integer n define $r_k^n = ktn^{-1}$ for $k = 0, 1, \ldots, n - 1$. If

$$w_n = \sum_{k=0}^{n-1} S(t - r_k^n)[f(r_k^n) - f(t)](r_{k+1}^n - r_k^n)$$

then $v_1(t) = \lim_{n \to \infty} w_n$. Since $t - r_k^n > 0$ for $k \le n - 1$ it follows that $w_n \in \mathcal{D}(A)$ and

$$Aw_n = \sum_{k=0}^{n-1} AS(t - r_k^n)[f(r_k^n) - f(t)](r_{k+1}^n - r_k^n).$$

Since

$$\|AS(t - s)[f(s) - f(t)]\| \le |t - s|^{-1}M \cdot M \cdot |s - t|^{\nu} = M^2|t - s|^{\nu - 1}$$

for $s \in [0, t[$, it follows that

$$\int_0^t AS(t - s)[f(s) - f(t)]\, ds = \lim_{n \to \infty} Aw_n$$

exists. Since A is closed, $v_1(t) \in \mathcal{D}(A)$ and

$$Av_1(t) = \int_0^t AS(t-s)\,[f(s) - f(t)]\,ds \quad \text{for all } t \in [0, T].$$

Thus, it remains to show that $t \mapsto \int_0^t AS(t-s)[f(s) - f(t)]\,ds$ is continuous on $[0, \infty[$. Since

$$\left\| \int_0^t AS(t-s)\,[f(s) - f(t)]\,ds \right\| \leq \int_0^t M^2\,|t-s|^{\nu-1}\,ds \to 0$$

as $t \to 0^+$, we have that $t \mapsto Av_1(t)$ is continuous at $t = 0$. Now let $t > 0$ and $\varepsilon > 0$. Choosing $\delta \in {]}0, t/2{[}$ so that

$$\left\| \int_{r-\delta}^r AS(r-s)\,[f(s) - f(r)]\,ds \right\| \leq \int_{r-\delta}^r M^2 |r-s|^{\nu-1}\,ds \leq \varepsilon/2$$

for $r \in [t/2, 2t]$ and noting that $\psi_\delta(r) \equiv \int_0^{r-\delta} AS(r-s)[f(s) - f(r)]\,ds$ is continuous at $r = t$, we see that

$$\lim_{r \to t} \|Av_1(r) - Av_1(t)\| \leq \varepsilon + \lim_{r \to t} \|\psi_\delta(r) - \psi_\delta(t)\| = \varepsilon.$$

Since this holds for each $\varepsilon > 0$ we have that $t \mapsto Av_1(t)$ is continuous. $\qquad\square$

1.5. Nonlinear differential equations on Banach spaces

This section deals with the existence and uniqueness of classical and strong solutions for the Cauchy problem on the space X,

$$\begin{cases} \dfrac{du}{dt} = A(t)u(t) \\ u(0) = u_0 \in \mathcal{D}(A(t)) \, , \end{cases} \tag{1.20}$$

where $A(t)$ is a nonlinear operator whose domain is independent of t, $\mathcal{D}(A(t)) \equiv \mathcal{D}$.

The following lemma is well–known (see e.g. Kato [1967, Lemma 1.3]):

Lemma 1.14. *Let u be an X–valued function on an interval of real numbers. Suppose u has a weak derivative $u'_w(s) \in X$ at $t = s$ (that is, $df(u(t))/dt$ exists at $t = s$ and equals $f(u'_w(s))$ for every $f \in X^*$). If $\|u(\cdot)\|$ is also differentiable at $t = s$, then*

$$\frac{d\|u(s)\|}{ds} = \operatorname{Re} g\,(u'_w(s))$$

for every $g \in G(u(s))$.

Proof. Since $\operatorname{Re} g(u(t)) \leq |g(u(t))| \leq \|u(t)\|$ and $\operatorname{Re} g(u(s)) = \|u(s)\|$ we have

$$\operatorname{Re} g(u(t) - u(s)) \leq \|u(t)\| - \|u(s)\|.$$

Dividing both sides by $t - s$ and letting $t \to s$ from above, we obtain $\operatorname{Re} g(u'_w(s)) \leq (d/ds)\|u(s)\|$. Letting $t \to s$ from below we obtain the reverse inequality. $\square$

The usefulness of the duality type functionals depends mainly on the following lemma (Coppel [1965]):

Lemma 1.15. *Let I be a real interval and $f : I \mapsto X$ such that $d^- f(t)/dt$ exists for $t \in I$. Then, $d^- \|f(t)\|/dt$ exists and*

$$\frac{d^- \|f(t)\|}{dt} = \left\langle f(t), \frac{d^- f(t)}{dt} \right\rangle_- .$$

There is a corresponding result for the right–hand derivatives.

Proof. It is enough to observe that for $h > 0$

$$\left| \frac{\|f(t)\| - \|f(t-h)\|}{h} - \frac{\|f(t) - h d^- f(t)/dt\| - \|f(t)\|}{-h} \right|$$

$$= \left| \frac{\|f(t-h)\| - \|f(t) - h d^- f(t)/dt\|}{-h} \right|$$

$$\leq \left\| \frac{f(t) - f(t-h)}{h} - \frac{d^- f(t)}{dt} \right\| \to 0 \text{ as } h \to 0^+.$$

$\square$

The following result of this section refers to the Cauchy problem (1.20) with an everywhere defined and continuous operator A. This theorem was independently obtained by Lovelady and Martin [1972] and Pavel [1972 a,b]. See also Pavel [1984, p. 65].

Theorem 1.8. *Suppose that the function $(t, x) \mapsto A(t)x$, $[0, +\infty[\times X \mapsto X$, is continuous and that there is a continuous function $c : [0, +\infty[\mapsto \mathbf{R}$ such that for each $t \geq 0$, $A(t) - c(t)\mathcal{I}$ is dissipative. Then, for each $u_0 \in X$, the problem (1.20) has a unique classical solution. Furthermore,*

$$\|u(t) - u_0\| \leq \int_0^t \|A(s)u_0\| \exp\left(\int_s^t c(r)\, dr \right)\, ds$$

and

$$\|u_1(t) - u_2(t)\| \le \|u_{10} - u_{20}\| \exp\left(\int_0^t c(s)\, ds\right) \quad \text{for all } t \ge 0,$$

where u_i is the solution of (1.20) with $u_i(0) = u_{i0} \in X$, $i = 1, 2$.

Proof. If we define

$$A_1(t)x = \exp\left(-\int_0^t c(s)\, ds\right) A(t)x \exp\left(\int_0^t c(s)\, ds\right)$$

and

$$A_2(t)x = A_1(t)x - c(t)x \ , \ t \ge 0 \ , \ x \in X,$$

it is easy to check that $A_2(t)$ is dissipative.

If we suppose that the result of the theorem in the case $c(t) \equiv 0$ is proved, then there is a unique function $v : [0, +\infty[\mapsto X$ satisfying

$$\begin{cases} v'(t) = A_2(t)v(t) \ , \\ v(0) = u_0. \end{cases}$$

It follows that the function $u : [0, +\infty[\mapsto X$ defined by

$$u(t) = v(t) \exp\left(\int_0^t c(s)\, ds\right)$$

is a solution of the problem (1.20). Therefore, it is enough to prove the theorem in the case $c(t) \equiv 0$. Moreover, the case $u_0 = 0$ is also sufficient to be proved.

Fix $T > 0$ and let $C([0, T] \ ; \ X) \equiv C$ be the space of all continuous functions $u : [0, T] \mapsto X$ with the usual "supremum" norm $\| \cdot \|_C$. Define the operator $U : C \to C$ by

$$(Uu)(t) = A(t)u(t) \ , \ u \in C \ , \ t \in [0, T].$$

The dissipativity of $A(t)$ implies (see **D5**):

$$\|u(t) - v(t)\| \le \|u(t) - v(t) - \lambda\,[A(t)u(t) - A(t)v(t)]\|$$

for all $\lambda > 0$, $u, v \in C$ and $t \in [0, T]$. Therefore,

$$\|u - v\|_C \le \|u - v - \lambda(Uu - Uv)\|_C \quad \text{for all } \lambda > 0.$$

Thus $U : C \to C$ is continuous and dissipative.

We now define the linear operator $B : \mathcal{D}(B) \subset C \mapsto C$ by

$$(Bu)(t) = -u'(t),$$

with $\mathcal{D}(B) = \{u \in C \; ; \; u' \in C \, , \; u(0) = 0\}$ where u' denotes the strong derivative of u. It is known that B is "m"–dissipative.

According to a theorem of Webb [1972], $U + B$ is "m"–dissipative too, i.e. for each $\lambda > 0$ there is $u_\lambda \in \mathcal{D}(B)$ such that

$$\begin{cases} \lambda u_\lambda(t) + u'_\lambda(t) = A(t)u_\lambda(t) \, , \\ u_\lambda(0) = 0 \, . \end{cases} \tag{1.21}$$

Let $g \in G(u_\lambda(t))$ be such that

$$\operatorname{Re} g\left(A(t)u_\lambda(t) - A(t)0\right) \le 0.$$

Then by (1.21) we easily derive

$$\operatorname{Re} g\left(u'_\lambda(t)\right) \le \operatorname{Re} g\left(A(t)0\right) - \lambda\|u_\lambda(t)\| \le \|A(t)0\|.$$

Hence, by Lemma 1.14

$$\frac{d}{dt}\|u_\lambda(t)\| \le \|A(t)0\| \quad \text{a.e. on } [0, T].$$

This implies $\|u_\lambda(t)\| \le TM$ for all $t \in [0, T]$ and $\lambda > 0$, where $M = \sup\{\|A(t)0\|;\ t \in [0, T]\}$. Returning to (1.21), we see that

$$u'_\lambda(t) - u'_\mu(t) = A(t)u_\lambda(t) - A(t)u_\mu(t) + \mu u_\mu(t) - \lambda u_\lambda(t) \, ,$$

which implies (a.e. on $[0, T]$)

$$\frac{d}{dt}\|u_\lambda(t) - u_\mu(t)\| \le \|\mu u_\mu(t) - \lambda u_\lambda(t)\| \le (\lambda + \mu)MT.$$

Since $u_\lambda(0) = u_\mu(0) = 0$, this inequality yields

$$\|u_\lambda(t) - u_\mu(t)\| \le (\lambda + \mu)MT^2 \, , \quad \text{for } \lambda, \mu > 0.$$

Hence $\lim_{\lambda \to 0+} u_\lambda(t) = u(t)$ exists uniformly on $[0, T]$. Letting $\lambda \to 0^+$ in (1.21) we obtain

$$\begin{cases} u'(t) = A(t)u(t) \quad \text{for all } t \in [0, T] \\ u(0) = 0. \end{cases}$$

On the other hand by Lemma 1.15 and dissipativity of A we derive:

$$\begin{aligned} \frac{d^-}{dt}\|u(t) - u_0\| = \langle u(t) - u_0, A(t)u(t)\rangle_- &\le \\ &\le \langle u(t) - u_0, A(t)u(t) - A(t)u_0\rangle_- + \|A(t)u_0\| \le \\ &\le c(t)\|u(t) - u_0\| + \|A(t)u_0\|. \end{aligned}$$

Solving this differential inequality we obtain the desired result. The second inequality of the theorem can be proved in a similar way. □

Let us consider now the operator $A : \mathcal{D}(A) \mapsto X$ and a real number c such that $A - c\mathcal{I}$ is "m"-dissipative. Following Proposition 1.3 we can define, for each positive integer n, the resolvent of $A - c\mathcal{I}$, namely

$$J_n = \left[\mathcal{I} - n^{-1}(A - c\mathcal{I}) \right]^{-1} : X \mapsto \mathcal{D}(A).$$

Also, let us consider the "Yosida approximation" of $-(A - c\mathcal{I})$ i.e. the everywhere defined operator

$$A_n = -(A - c\mathcal{I})J_n = n(\mathcal{I} - J_n).$$

Finally, let $B_n : X \mapsto X$ be the operator

$$B_n = AJ_n = -A_n + cJ_n = -n\mathcal{I} + (n + c)J_n.$$

Let us list several well–known properties of these operators. The proofs can be found, for instance in Pavel [1984, pp. 20–22].

Lemma 1.16. *If $A - c\mathcal{I}$ is "m"-dissipative, then:*

 i) $\|J_n x - J_n y\| \leq \|x - y\|$ *for all* $x, y \in X$,
 ii) $\|A_n x - A_n y\| \leq 2n\|x - y\|$ *for all* $x, y \in X$,
 iii) $\langle x - y, -(A_n x - A_n y)\rangle_+ \leq 0$ *for all* $x, y \in X$, i.e. *the Yosida approximation is totally dissipative*,
 iv) $\|B_n x - B_n y\| \leq (2n + |c|)\|x - y\|$ *for all* $x, y \in X$,
 v) $\langle x - y, B_n x - B_n y\rangle_+ \leq |c|\|x - y\|$ *for all* $x, y \in X$,
 vi) *if* $x \in \mathcal{D}(A)$,

$$\|A_n x\| \leq \|(A - c\mathcal{I})x\| \quad \text{and}$$

$$\|B_n x\| \leq (1 + |c|n^{-1})\|(A - c\mathcal{I})x\| + \|cx\|,$$

 vii) *if* $x \in \overline{\mathcal{D}(A)}$, $J_n x \to x$ *as* $n \to \infty$.

□

Other properties of these operators are given in the following result:

Lemma 1.17. *Let $A - c\mathcal{I}$ be "m"-dissipative and suppose that for each sequence $\{x_n\}_n$ in $\mathcal{D}(A)$ such that $x_n \to x$ and that $\|Ax_n\|$ are bounded, it follows that $x \in \mathcal{D}(A)$ and $Ax_n \overset{w}{\to} Ax$. Then:*

 i) *If $\{y_n\}_n$ is a sequence in X such that $y_n \to y$ and that $\|A_n y_n\|$ are bounded, then $y \in \mathcal{D}(A)$, $A_n y_n \overset{w}{\to} -(A - c\mathcal{I})y$ and $B_n y_n \overset{w}{\to} Ay$.*
 ii) *If z is in $\mathcal{D}(A)$ then $A_n z \overset{w}{\to} -(A - c\mathcal{I})z$ and $B_n z \overset{w}{\to} Az$.*

Proof. Letting $x_n = J_n y_n \in \mathcal{D}(A)$ we have $y_n - x_n = n^{-1} A_n y_n \to 0$ so that $x_n \to y$. On the other hand,

$$\|A x_n\| = \|A J_n y_n\| \leq \|A_n y_n\| + |c| \|x_n\|$$

i.e. the $\|A x_n\|$ are bounded. This means that $y \in \mathcal{D}(A)$ and $A x_n \xrightarrow{w} Ay$ i.e. $A_n y_n = -A x_n + c x_n \xrightarrow{w} -(A - cI)y$. Consequently $B_n y_n \xrightarrow{w} Ay$. Thus i) is true. Part ii) follows from i) with $y_n = z$ and Lemma 1.16 vi). $\qquad \square$

Let us consider now a family of operators with the same domain $\mathcal{D}(A(t)) = \mathcal{D}$, $\{A(t) \, ; \, A(t) : \mathcal{D} \mapsto X \, , \, t \geq 0\}$ having one of the properties:

H1. There is a continuously differentiable function $c : [0, \infty[\mapsto \mathbf{R}$ such that $A(t) - c(t)I$ is "m"-dissipative for all $t \geq 0$.

H2. There is a continuous function $d : ([0, \infty[)^3 \mapsto [0, \infty[$ such that

$$\|A(t)x - A(s)x\| \leq |t - s| d(t, s, \|x\|) (1 + \|A(t)x\| + \|A(s)x\|)$$

for all $t, s \geq 0$ and all x in $\mathcal{D}$.

H3. If $t \geq 0$ and $\{x_n\}_n$ is a sequence in $\mathcal{D}$ such that $x_n \to x$ and $\|A(t)x_n\|$ are bounded for $n \geq 1$, then $x \in \mathcal{D}$ and $A(t)x_n \xrightarrow{w} A(t)x$.

Lemma 1.18. *If **H2** is valid for each bounded subset $Q \subset \mathcal{D}$ and there is a $\delta > 0$ and an $M > 0$ such that if x is in Q and $t, s \in [0, T]$ with $|t - s| \leq \delta$, then*

$$\|A(t)x - A(s)x\| \leq |t - s| M (1 + 2\|A(s)x\|) .$$

Proof. Take $M = 2 \sup\{d(t, s, \|x\|) \, ; \, x \in Q, \, t, s \in [0, T]\}$ and $\delta = 1/M$. For $x \in Q$ and $|t - s| \leq \delta$ we obtain from **H2**:

$$\|A(t)x - A(s)x\| \leq |t - s| M (1 + \|A(t)x - A(s)x\| + 2\|A(s)x\|) / 2 \leq$$
$$\leq \|A(t)x - A(s)x\|/2 + |t - s| M (1 + 2\|A(s)x\|)/2$$

and the assertion of the lemma follows. $\qquad \square$

Lemma 1.19. *If **H1** and **H2** are fulfilled and Q is a bounded set of X, then there is a constant K such that $\|J_n(t)x\| \leq K$ for all (t, x) in $[0, T] \times Q$ and all $n \geq 1$.*

Proof. Let M be such that $\|x\| \leq M$ for all $x \in Q$, let $z \in \mathcal{D}$, and take $K = M + \sup\{\|A(t)z - c(t)z\| \, ; \, t \in [0, T]\} + 2\|z\|$ (see Lemma 1.18). By part i) of Lemma 1.16,

$$\|J_n(t)x\| \leq \|J_n(t)x - J_n(t)z\| + \|J_n(t)z\| \leq$$
$$\leq \|x - z\| + \|[I - n^{-1} A_n(t)]z\| \leq \|x\| + 2\|z\| + n^{-1} \|A_n(t)z\|.$$

The lemma now follows from vi) of Lemma 1.16. $\qquad \square$

Lemma 1.20. *If* **H1** *and* **H2** *hold, then* $(t, x) \mapsto B_n(t)x$ *is continuous from* $[0, \infty[\times X$ *into* X.

Proof. From i) of Lemma 1.16 we have

$$\|J_n(t)x - J_n(s)x\|$$
$$= \left\| J_n(t) \left[\mathcal{I} - n^{-1} \left(A(s) - c(s)\mathcal{I} \right) \right] J_n(s)x \right.$$
$$\left. - J_n(t) \left[\mathcal{I} - n^{-1} \left(A(t) - c(t)\mathcal{I} \right) \right] J_n(s)x \right\|$$
$$\leq n^{-1} \|A(t)J_n(s)x - A(s)J_n(s)x\| + n^{-1} |c(t) - c(s)| \, \|J_n(s)x\| .$$

Thus,

$$\|B_n(t)x - B_n(s)x\|$$
$$= \left\| [n + c(t)] [J_n(t)x - J_n(s)x] + [c(t) - c(s)] J_n(s)x \right\|$$
$$\leq |1 + n^{-1}c(t)| \, \|A(t)J_n(s)x - A(s)J_n(s)x\|$$
$$+ |1 + n^{-1}c(t)| |c(t) - c(s)| \, \|J_n(s)x\| .$$

If $t \in [0, T]$ and x is in a bounded set Q, from Lemmas 1.18 and 1.19, it follows that there is a $\delta > 0$ and constants M' and K such that if $|t - s| \leq \delta$, then

$$\|B_n(t)x - B_n(s)x\|$$
$$\leq |1 + n^{-1}c(t)| \, |t - s|M' \left(1 + 2\|A(s)J_n(s)x\| \right) + |1 + n^{-1}c(t)| \, |c(t) - c(s)|K.$$

Taking into account the continuous differentiability of c we conclude that for each bounded set Q, there exists a $\delta > 0$ and an $M > 0$ such that

$$\|B_n(t)x - B_n(s)x\| \leq |t - s|M \left(1 + 2\|B_n(s)x\| \right) \tag{1.22}$$

for all $n \geq 1$, x in Q and $t, s \in [0, T]$ with $|t - s| \leq \delta$. Moreover, Lemma 1.16 iv) gives

$$\|B_n(t)x - B_n(s)y\| \leq |t - s|M(1 + 2\|B_n(s)x\| + (2n + |c(t)|) \, \|x - y\|$$

which implies the desired continuity. $\qquad\qquad\qquad\qquad\qquad\qquad\square$

Now we are ready to state our main result in this section. Let us note that a similar theorem was proved by Martin [1970, Th. 4.1.]. Our result is a slight extension of that one, because our class of "m" totally dissipative operators (hypothesis **H1**) is larger than the class of "uniformly m–monotone" operators with which Martin works.

Theorem 1.9. *Let us consider the family of operators* $\{A(t)$; $A(t) : \mathcal{D} \mapsto X,$
$t \geq 0\}$ *restricted by assumptions* **H1, H2, H3** *and let* u_0 *be in* $\mathcal{D}$. *Then, there is
a unique function* $u : [0, \infty[\mapsto \mathcal{D}$ *with the following properties*

  i)   *u is Lipschitz continuous on bounded subintervals of $[0, \infty[$.*
  ii) *$u(0) = u_0$, the weak derivative u'_w of u exists, is weakly continuous, and
       satisfies $u'_w(t) = A(t)u(t)$ for all $t \geq 0$.*
  iii) *The derivative du/dt of u exists almost everywhere on $[0, \infty[$ and $du/dt =
       A(t)u(t)$ for almost all $t \geq 0$ (i.e. u is a strong solution of (1.20)).*
  iv) *If u_1 and u_2 are two strong solutions of (1.20) corresponding to initial
       conditions u_{10} and u_{20} respectively, then*

$$\|u_1(t) - u_2(t)\| \leq \|u_{10} - u_{20}\| \exp \int_0^t c(s)\, ds.$$

Proof. For each $n \geq 1$, let us consider the "approximate" Cauchy problem

$$\begin{cases} du_n/dt = B_n(t)u_n(t) \\ u_n(0) = u_0 \in \mathcal{D}. \end{cases} \tag{1.23}$$

Taking into account Lemma 1.20, Lemma 1.16 v) and Theorem 1.8 we deduce
the existence of the unique continuously differentiable function $u_n : [0, \infty[\mapsto \mathcal{D}$
satisfying (1.23) for all $t \geq 0$, and

$$\|u_n(t) - u_0\| \leq \int_0^t \|B_n(s)u_0\| \exp \left(\int_s^t |c(r)|\, dr \right) ds. \tag{1.24}$$

Due to Lemma 1.16 vi) we have the boundedness of $\{\|B_n(s)u_0\|\}_n$ and (1.24)
implies the existence of a constant K such that

$$\|u_n(t)\| \leq K \text{ for all } n \geq 1 \text{ and } t \in [0, T]. \tag{1.25}$$

Now let Q be a bounded subset of X which contains $u_n(t)$ for all $t \in [0, T]$ and
$n \geq 1$. Moreover, let $\delta > 0$ and $M > 0$ be such that (1.22) is valid.

Taking $0 < h \leq \delta$ and $t \in [0, T]$ and by using Lemma 1.15, Lemma 1.1 parts iv)
and vi), Lemma 1.16 v) and the inequality (1.22) we successively obtain:

$$\begin{aligned}
\frac{d^-}{dt} & \|u_n(t+h) - u_n(t)\| \\
&= \langle u_n(t+h) - u_n(t), B_n(t+h)u_n(t+h) - B_n(t)u_n(t) \rangle_- \\
&\leq \langle u_n(t+h) - u_n(t), B_n(t+h)u_n(t+h) - B_n(t+h)u_n(t) \rangle_- \\
&\quad + \|B_n(t+h)u_n(t) - B_n(t)u_n(t)\| \\
&\leq |c(t)|\, \|u_n(t+h) - u_n(t)\| + hM\left(1 + 2\|B_n(t)u_n(t)\|\right).
\end{aligned}$$

Consequently,

$$\|u_n(t+h) - u_n(t)\| \leq \|u_n(h) - u_n(0)\| \exp\left(\int_0^t |c(s)|\, ds\right) +$$

$$+ hM \int_0^t (1 + 2\|B_n(s) \cdot u_n(s)\|) \exp\left(\int_s^t |c(r)|\, dr\right)\, ds.$$

Dividing by h, letting $h \to 0^+$ and noting that $B_n(s)u_n(s) = du_n(s)/dt$, we have

$$\left\|\frac{du_n(t)}{dt}\right\| \leq \|B_n(0)u_0\| \exp\left(\int_0^t |c(s)|\, ds\right) +$$

$$+ M \int_0^t (1 + 2\|du_n(s)/ds\|) \exp\left(\int_s^t |c(r)|\, dr\right)\, ds.$$

Since $\|B_n(0)u_0\|$ is bounded by part vi) of Lemma 1.16, it follows from Gronwall's inequality (see e.g. Coppel [1965, p. 19]) that there is a constant K such that

$$\|du_n(t)/dt\| = \|B_n(t)u_n(t)\| \leq K \quad \text{for all } t \in [0, T] \text{ and } n \geq 1. \tag{1.26}$$

Let K_1 be a constant such that $|c(t)| \leq K_1$ for all $t \in [0, T]$. Let also ε be a positive number and $\varepsilon' = K_1 \varepsilon / 2(\exp(K_1 T) - 1)$.

Following again Lemma 1.15 and also Lemma 1.1 part xi), there is $h(\varepsilon) > 0$ such that

$$\frac{d^-}{dt}\|u_m(t) - u_n(t)\|$$

$$= \left\langle u_m(t) - u_n(t) + J_m(t)u_m(t) - J_n(t)u_n(t) - J_m(t)u_m(t)+ \right.$$

$$\left. + J_n(t)u_n(t), A(t)J_m(t)u_m(t) - A(t)J_n(t)u_n(t) \right\rangle_-$$

$$\leq \langle J_m(t)u_m(t) - J_n(t)u_n(t), A(t)J_m(t)u_m(t) - A(t)J_n(t)u_n(t) \rangle_- +$$

$$+ \|u_m(t) - J_m(t)u_m(t) - u_n(t) + J_n(t)u_n(t)\| \cdot h(\varepsilon) + \varepsilon',$$

where u_n and u_m are two classical solutions of (1.23). By using hypothesis **H1** and Lemma 1.16 i) we derive

$$\frac{d^-}{dt}\|u_m(t) - u_n(t)\| \leq K_1\|J_m(t)u_m(t) - J_n(t)u_n(t)\| +$$

$$+ \|u_m(t) - J_m(t)u_m(t)\| \cdot h(\varepsilon) + \|u_n(t) - J_n(t)u_n(t)\| \cdot h(\varepsilon) + \varepsilon'$$

$$\leq (K_1 + h(\varepsilon))\left(\|u_m(t) - J_m(t)u_m(t)\| + \|u_n(t) - J_n(t)u_n(t)\|\right) +$$

$$+ K_1\|u_m(t) - u_n(t)\| + \varepsilon'.$$

But,

$$\|u_n(s) - J_n(s)u_n(s)\| = n^{-1}\|A_n(s)u_n(s)\|$$
$$\leq n^{-1}\|B_n(s)u_n(s)\| + n^{-1}\|c(s)J_n(s)u_n(s)\| \leq n^{-1}K_2$$

as (1.26), (1.25) and Lemma 1.16 i) give.

Then,

$$\frac{d^-}{dt}\|u_m(t) - u_n(t)\|$$
$$\leq (K_1 + h(\varepsilon))\,K_2(n^{-1} + m^{-1}) + K_1\|u_m(t) - u_n(t)\| + \varepsilon'$$
$$\leq K_1\|u_m(t) - u_n(t)\| + K_1\varepsilon/(\exp K_1 T - 1)$$

for all $m, n \geq n_0 = 4K_2[K_1 + h(\varepsilon)](\exp(K_1 T) - 1)/K_1\varepsilon$. Hence, the differential inequality implies $\|u_m(t) - u_n(t)\| \leq \varepsilon$, whenever $m, n \geq n_0$ and $t \in [0, T]$. Consequently, the sequence $\{u_n\}_n$ is uniformly Cauchy and since X is complete, there is a continuous function $u : [0, T] \mapsto X$ such that $u_n(t) \to u(t)$ uniformly on $[0, T]$.

As $\|du_n/dt\|$ are bounded for $t \in [0, T]$ and $n \geq 1$, it follows that u is Lipschitz continuous on $[0, T]$. On the other hand, since $u_n(t) \to u(t)$ and $\|B_n(t)u_n(t)\| \leq K$, we obtain by Lemma 1.17 and the assumption **H3** : $\|A_n(t)u_n(t)\|$ are bounded and $u(t)$ is in $\mathcal{D}$, $B_n(t)u_n(t) \xrightarrow{w} A(t)u(t)$ and $\|A(t)u(t)\| \leq K$.

Now let Q be the bounded set $Q = \{u(t) \; ; \; t \in [0, T]\} \subset X$. By Lemma 1.18, let δ and M such that

$$\|A(t)u(t) - A(s)u(t)\| \leq |t - s|M(1 + 2K)$$

whenever $|t - s| \leq \delta$. Furthermore, since $u(t) \to u(s)$ as $t \to s$, we have by assumption **H3** that $A(s)u(t) \xrightarrow{w} A(s)u(s)$. Hence,

$$A(t)u(t) - A(s)u(s) = A(t)u(t) - A(s)u(t) + A(s)u(t) - A(s)u(s) \xrightarrow{w} 0$$

and it follows that $t \mapsto A(t)u(t)$ is weakly continuous on $[0, T]$. By (1.23) we obtain for all $f \in X^*$,

$$f(u_n(t)) = f(u_0) + \int_0^t f(B_n(s)u_n(s))\ ds.$$

Since $u_n(t) \to u(t)$, $B_n(s)u_n(s) \xrightarrow{w} A(t)u(t)$ and

$$\|f(B_n(s)u_n(s))\| \leq K\|f\|$$

we have (Lebesgue Dominated Convergence theorem),

$$f(u(t)) = f(u_0) + \int_0^t f(A(s)u(s))\ ds$$

and due to the weak continuity of $A(\cdot)u(\cdot)$ we obtain

$$\frac{df(u(t))}{dt} = f(A(t)u(t)) \text{ for all } t \in [0,T].$$

So we have proved parts i) and ii) of the theorem. As regards part iii) it is sufficient to prove that $t \mapsto A(t)u(t)$ is Bochner integrable on $[0,T]$ and $u(t) = u_0 + \int_0^t A(s)u(s)\,ds$. For the proof, the reader can see Kato [1967, Lemma 4.6]. For part iv) of the theorem we simply apply Lemma 1.15 and take into account the hypothesis **H1**.   □

Chapter II
Lumped parameter approach of
nonlinear networks with transistors

2.0. Introduction

In this chapter we study the lumped parameter modelling of a large class of circuits composed of bipolar transistors, junction diodes and passive elements (resistors, capacitors, inductors). All these elements are nonlinear: the semiconductor components are modelled by "large signal" equivalent schemes, the capacitors and inductors have monotone characteristics while the resistors can be included in a multiport which also has a monotone description.

In Section 2.1 we state the equations which describe the dynamic and direct current (steady state) behavior. The hypotheses are also formulated.

The core of the chapter is Section 2.2 where we show that the mathematical model contains a dissipative operator on $\mathbf{R}^N$ with remarkable properties.

Starting from here, in Sections 2.3 and 2.4 we infer qualitative properties of the model: the existence and uniqueness of steady state and dynamic solutions, their boundedness, stability and source dependence. The asymptotic behavior of the solution can be evaluated by bordering "delay time" with easily computable limits. An example is given in Section 2.5.

The interest in qualitative study of this class of circuits began in the late 1960's with papers of Sandberg and Willson. The reader can find in Marinov [1990 b] a list of references on this subject together with some comments.

The results of this chapter are mainly an extension of the results in Sandberg [1969]. The generalization consists, firstly, of considering a nonlinear resistive multiport of continuous piecewise continuously differentiable type (CPWCD), instead of the continuous linear one in Sandberg's work. The restrictions imposed upon nonlinearities (see assumptions **IV** and **V** or **IV*** and **V*** below) are related to the uniform diagonal dominance of the Jacobian matrix and therefore constitute a natural extension of the hypotheses in Sandberg [1969]. A second (partial) generalization consists of the description of all nonlinear elements outside the resistive multiport – the capacitors and inductors connected at its ports as well as the resistors included in transistor models – by continuous piecewise linear functions (CPWL) instead of the C^1 description considered by Sandberg. As regards our results about DC solutions presented in Section 2.3, the main achievement of this section is a partial extension of some known properties valid for the linear or piecewise linear multiport of the CPWCD case.

2.1. Mathematical model

Before describing the circuit under study, some prerequisites are needed. The $\mathbf{R}^N$ norm used below will be a weighted ℓ^1 one, namely $\|x\|_d = \sum_{i=1}^N d_i |x_i|$ where $d_1, d_2, ..., d_N$ are strictly positive constants. If we denote by $\|| \cdot \||_d$ the matrix norm induced by $\| \cdot \|_d$, and consider "the measure" of the $N \times N$ matrix M, namely

$$\mu_d(M) = \lim_{h \to 0+} \frac{\||\mathcal{I} + hM\||_d - 1}{h} ,$$

then (see Coppel [1965]) we have

$$\mu_d(M) = \max_{k=1,...,N} \left(M_{kk} + \sum_{\substack{i=1 \\ i \neq k}}^N \frac{d_i}{d_k} |M_{ik}| \right) . \tag{2.1}$$

The utility of this concept for the dissipativity on $\mathbf{R}^N$, follows from the following (almost straightforward) inequalities:

$$-\mu_d(-M)\|a\|_d \leq \langle a, Ma \rangle_- \leq \langle a, Ma \rangle_+ \leq \mu_d(M)\|a\|_d \tag{2.2}$$

where $a \in \mathbf{R}^N$.

Also, if $f : [0,1] \mapsto \mathbf{R}^N$ is continuous, then

$$\int_0^1 \langle a, f(\lambda) \rangle_- \, d\lambda \leq \langle a, \int_0^1 f(\lambda) \, d\lambda \rangle_- \leq \langle a, \int_0^1 f(\lambda) \, d\lambda \rangle_+ \leq \int_0^1 \langle a, f(\lambda) \rangle_+ \, d\lambda \tag{2.3}$$

where the Riemann integrals are taken componentwise. The proof can be found in Marinov [1990 b].

If $D \subset \mathbf{R}^N$ is a convex set and $F : D \mapsto \mathbf{R}^N$ has its Fréchet derivative F' continuous in D, then for any $x, y \in D$,

$$F(y) - F(x) = \int_0^1 F'(x + t(y - x))(y - x) \, dt \tag{2.4}$$

where the integral is also in componentwise sense (see, for example, Ortega and Rheinboldt [1970]).

Finally some definitions are necessary. We define hyperplanes P_i, $i = 1, ..., p$ in $\mathbf{R}^N$:

$$P_i = \left\{ x \in \mathbf{R}^N \mid \sum_{k=1}^N c_k^i x_k = b^i; \quad c_k^i, \, b^i \in \mathbf{R} \right\}$$

Let us consider the decomposition $\mathbf{R}^N - \cup_{i=1}^{p} P_i = \cup_{i=1}^{q} R_i$, where R_i, $i = 1, ..., q \le 2^p$ are disjoint, open and convex sets (called "regions") with a boundary contained in $\cup_{i=1}^{p} P_i$.

The function $F : \mathbf{R}^N \mapsto \mathbf{R}^N$ is called "continuous piecewise continuously differentiable" – CPWCD – if it is continuous everywhere and in any region R_i the (Fréchet) derivative F' exists and is continuous.

The function $F : \mathbf{R}^N \mapsto \mathbf{R}^N$ is called "continuous piecewise linear" – CPWL – if $F(x) = A_i x + B_i$ for all $x \in \overline{R_i}$. The Figure 2.1 presents the types of nonlinearities (in the scalar case) which appear in this chapter.

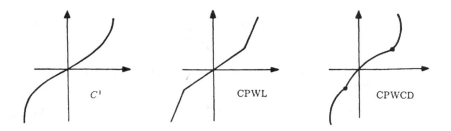

Figure 2.1 Different types of nonlinearities in $\mathbf{R}^1$

Let us consider the nonlinear network of Figure 2.2. The box represents a resistive nonlinear multiport with independent or/and controlled sources. To its N ports are connected P bipolar (npn and/or pnp) transistors, Q junction diodes, R nonlinear capacitors and S nonlinear inductors, so we have $N = 2P + Q + R + S$ components.

The semiconductor devices are described by the dynamic response large signal model proposed by Gummel [1968] and presented in Figure 2.3 (a) for the jth transistor and in Figure 2.3 (b) for the kth diode. These models take into account the nonlinear DC properties as well as the presence of nonlinear junction capacitances. They include standard circuit elements with six constant parameters α_f^j, α_r^j, τ_{2j-1}, τ_{2j}, c_{2j-1} and c_{2j} having strict positive values and $\alpha_f^j < 1$, $\alpha_r^j < 1$, and two nonlinear functions f_{2j-1}, f_{2j}. The resistances of conducting layers are included in the resistive multiport.

The following notation is used for the current vector

$$i = \begin{bmatrix} i^1 \\ i^2 \end{bmatrix}$$

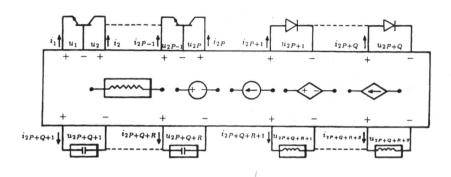

Figure 2.2 The circuit under study.

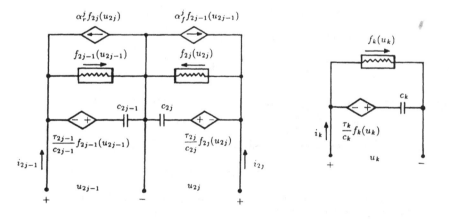

$$(a) \qquad\qquad\qquad\qquad\qquad (b)$$

Figure 2.3 The semiconductor device models

where

$$i^1 = [i_1, i_2, ..., i_{2P}, i_{2P+1}, ..., i_{2P+Q}, i_{2P+Q+1}, ..., i_{2P+Q+R}]^{tr}$$
$$i^2 = [i_{2P+Q+R+1}, ..., i_N]^{tr}$$

and analogously for the voltage vector u. Also, the state variable will be written as

$$z = \begin{bmatrix} q \\ \psi \end{bmatrix}$$

where

$$q = [q_1, ..., q_{2P}, q_{2P+1}, ..., q_{2P+Q}, q_{2P+Q+1}, ..., q_{2P+Q+R}]^{tr}$$

is the vector of capacitor charges (both in semiconductor device models and exterior capacitors) and

$$\psi = [\psi_1, ..., \psi_S]^{tr}$$

is the flux linkage vector in inductors. We also put

$$v = \begin{bmatrix} u^1 \\ i^2 \end{bmatrix}, \quad w = \begin{bmatrix} i^1 \\ u^2 \end{bmatrix}.$$

It is elementary to show that

$$w = TF(v) + \frac{d}{dt}z \tag{2.5}$$

where we have denoted $F: \mathbf{R}^N \mapsto \mathbf{R}^N$ by

$$F(x) = [f_1(x_1), ..., f_{2P+Q}(x_{2P+Q}), 0, ..., 0]^{tr}$$

$$T = \bigoplus_{j=1}^{P} \begin{bmatrix} 1 & -\alpha_r^j \\ -a_f^j & 1 \end{bmatrix} \oplus I_Q \oplus 0_{R+S}.$$

Here I_Q is the $Q \times Q$ identity matrix, 0_{R+S} is a zero matrix with $R + S$ rows and columns, $A \oplus B$ means $\begin{bmatrix} A & 0 \\ 0 & B \end{bmatrix}$ and x_j is the jth component of x.

From the above models we also easily derive

$$q_j \overset{\text{def}}{=} \gamma_j(u_j) = c_j u_j + \tau_j f_j(u_j)$$

for $j = 1, 2, ..., 2P + Q$. We suppose that the extra device capacitors and inductors are described by the nonlinear functions γ_j, i.e. $q_j = \gamma_j(u_j)$ for $j = 2P + Q + 1, ..., 2P + Q + R$ and $\psi_j = \gamma_j(i_j)$ for $j = 2P + Q + R + 1, ..., N$.

Our first hypothesis regards the scalar nonlinearities.

I. (a) For every $j = 1, 2, ..., 2P + Q$ the function $f_j : \mathbf{R} \mapsto \mathbf{R}$ is CPWL and in every region (here interval) $f_j'(x) > 0$.

   (b) For every $j = 2P + Q + 1, ..., N$ the function $\gamma_j : \mathbf{R} \mapsto \mathbf{R}$ is CPWL and there exist strictly positive constants η_j and ξ_j such that $0 < \eta_j \le \gamma_j'(x) \le \xi_j$ for any x in any region (interval).

The second assumption is related to the "hybrid" description of the nonlinear resistive multiport.

II. (a) There exists $H : \mathbf{R}^N \mapsto \mathbf{R}^N$ and $B : [0, \infty[\mapsto \mathbf{R}^N$ such that $w = -H(v) + B(t)$.

    (b) H is CPWCD.

    (c) B is continuous.

We shall use hypotheses III and/or IV and/or V regarding the Jacobian matrix of H.

III. There exist strictly positive numbers $d_1, ..., d_N$ such that for every $j = 1, ..., N$ there is an $\omega_j \in \mathbf{R}$ for which in any region the following holds:

$$-\frac{\partial H_j}{\partial x_j}(x) + \sum_{\substack{i=1 \\ i \neq j}}^{N} \frac{d_i}{d_j} \left| \frac{\partial H_i}{\partial x_j}(x) \right| \leq \omega_j \ . \tag{2.6}$$

IV. There exist strictly positive numbers $d_1, ..., d_N$ such that for all $k = 1, ..., P$

$$\alpha_f^k < \frac{d_{2k-1}}{d_{2k}} < \frac{1}{\alpha_r^k} \tag{2.7}$$

and for every $j = 1, ...N$ there is an $\omega_j < 0$ satisfying inequality (2.6) in any region.

V. There exist strictly positive numbers $d_1, ..., d_N$ such that for every $j = 1, ..., N$ there is a $\beta_j < 0$ which in any region makes true the inequality

$$-\frac{\partial H_j}{\partial x_j}(x) - \sum_{\substack{i=1 \\ i \neq j}}^{N} \frac{d_i}{d_j} \left| \frac{\partial H_i}{\partial x_j}(x) \right| \geq \beta_j \ . \tag{2.8}$$

Now we shall formulate a new series of hypotheses, parallel with the previous ones but with smoother functions.

I*. (a) For every $j = 1, 2, ..., 2P + Q$ the function $f_j : \mathbf{R} \mapsto \mathbf{R}$ is of C^1 type and $f_j'(x) > 0$ for $x \in \mathbf{R}$.

    (b) For every $j = 2P + Q + 1, ..., N$ the function $\gamma_j : \mathbf{R} \mapsto \mathbf{R}$ is of C^1-type and $0 < \eta_j \leq \gamma_j'(x) \leq \xi_j$ for $x \in \mathbf{R}$.

II*. The same as **II** but replacing the CPWCD property by the C^1 property of H.

III*, IV*, V*. The same as **III, IV, V** respectively but using "for all $x \in \mathbf{R}^N$" instead of "in any region".

All our results will be stated supposing either (some of) assumptions **I–V** or (some of) assumptions **I*–V***. The proofs will be given only for the first case, their transposition to the second being obvious.

If hypothesis **I** is valid, then we observe that for all $j = 1, ..., N$ the functions γ_j are invertible and $\gamma_j^{-1} = g$ is CPWL. Then we can define the function $G : \mathbf{R}^N \mapsto \mathbf{R}^N$ by

$$G(x) = [g_1(x_1), ..., g_N(x_N)]^{tr}$$

which has the CPWL property in $\mathbf{R}^N$. Similarly, if **I*** is valid, $\gamma_j^{-1} = g_j$ exists for all j and is of C^1 class. Also we can define as above $G : \mathbf{R}^N \mapsto \mathbf{R}^N$, $G \in C^1$.

With the previous notation we have $G(z) = v$ and we may define the function $A : \mathbf{R}^N \mapsto \mathbf{R}^N$ by

$$A(z) = -TF(G(z)) - H(G(z)) . \tag{2.9}$$

If, in addition, hypothesis **II** (a) holds, then from (2.5) and (2.9) we derive the following differential equation in $\mathbf{R}^N$ (with associated initial condition) describing the dynamic behavior of our network:

$$(E(z_0, A, B)) \qquad \begin{cases} \dfrac{dz}{dt} = A(z) + B(t) \\ z(0) = z_0 \in \mathbf{R}^N, \quad t \geq 0 . \end{cases}$$

Corresponding to this problem we can formulate the steady state (or DC) equation of the network under study:

$$(S(A, \overline{B})) \qquad \begin{cases} A(z) + \overline{B} = 0 \\ \overline{B} \in \mathbf{R}^N . \end{cases}$$

2.2. Dissipativity

If we accept the hypotheses I and II, let us denote by $R(r_i)$, $r_i = 1, ..., p_i$, the open regions (intervals) defining the CPWL structure of g_i, $i = 1, ..., N$. That is $\mathbf{R} = \cup_{r_i=1}^{p_i} R(r_i)$ and $g_i(x_i) = m^{r_i} x_i + n^{r_i}$ for all $x_i \in \overline{R(r_i)}$. Let also, for $k = 1, ..., p$, $P_k(x) = \{x \in \mathbf{R}^N \mid \sum_{j=1}^{N} c_j^k x_j = b^k\}$ be hyperplanes in $\mathbf{R}^N$ on which H is not differentiable. Let us denote

$$M_1 = \{x \in \mathbf{R}^N \mid H \circ G \text{ is not differentiable in } x\}$$
$$M_2 = \{x \in \mathbf{R}^N \mid G \text{ is not differentiable in } x\}$$
$$M_3 = \{x \in \mathbf{R}^N \mid H \text{ is not differentiable in } G(x)\}$$
$$M_4 = \bigcup_{k=1}^{p} \bigcup_{r_1=1}^{p_1} ... \bigcup_{r_N=1}^{p_N} \left\{x \in \mathbf{R}^N \,\middle|\, \sum_{j=1}^{N} c_j^k(m^{r_j} x_j + n^{r_j}) = b^k\right\} .$$

Taking into account that $M_1 \subset M_2 \cup M_3 \subset M_2 \cup M_4$ we observe that $H \circ G$ is CPWCD in respect to $M_2 \cup M_4$, a union of hyperplanes. Also, we easily observe that $F \circ G$ is CPWL. Therefore, we conclude that the hypotheses **I** and **II** imply the CPWCD property of A. In addition, **I*** and **II*** imply the C^1 property of A.

Starting from these remarks we can prove our first result, which will be essential below.

Lemma 2.1. *Let the hypotheses* **I, II, III** *(or* **I*, II*, III***) *be valid. Then,*

(a) *there exists $\omega \in \mathbf{R}$ such that, for any region of A and for any z there (respectively, for any $z \in \mathbf{R}^N$) we have*

$$\mu_D(A'(z)) \leq \omega ,$$

(b) $A - \omega \mathcal{I}$ *is totally dissipative in* $\mathbf{R}^N$.

Proof. (a) Let us consider $j \in \{1, ..., 2P + Q\}$ and $d_1, .., d_N$ the numbers from the assumption **III**. Denoting by A_i the components of A and by t_{ij} the elements of T we have:

$$\frac{\partial A_j}{\partial z_j}(z) + \sum_{\substack{i=1 \\ j \neq j}}^{N} \frac{d_i}{d_j} \left| \frac{\partial A_i}{\partial z_j} \right| = \left[-t_{jj} f_j'(g_j(z_j)) - \frac{\partial H_j}{\partial x_j}(G(z)) + \right.$$

$$\sum_{\substack{i=1 \\ i \neq j}}^{N} \frac{d_i}{d_j} \left| t_{ij} f_j'(g_j(z_j)) + \frac{\partial H_i}{\partial x_j}(G(z)) \right| \left] \Big/ \left[c_j + \tau_j f_j'(g_j(z_j)) \right] \leq \qquad (2.10)$$

$$\max\left[\left(-t_{jj} + \sum_{\substack{i=1 \\ i \neq j}}^{N} \frac{d_i}{d_j} |t_{ij}| \right) \Big/ \tau_j \; ; \; \omega_j/c_j \right] = \max(s_j/\tau_j \; ; \; \omega_j/c_j)$$

where we have denoted

$$s_j = -t_{jj} + \sum_{\substack{i=1 \\ i \neq j}}^{N} |t_{ij}| = \begin{cases} -1 + \dfrac{d_{j+1}}{d_j} \alpha_f^{(j+1)/2} & \text{for } j \in \{1, 3, ..., 2P-1\} \\[2mm] -1 + \dfrac{d_{j-1}}{d_j} \alpha_r^{j/2} & \text{for } j \in \{2, 4, ..., 2P\} \\[2mm] -1 & \text{for } j \in \{2P+1, ..., 2P+Q\} . \end{cases} \qquad (2.11)$$

For the same numbers $d_1, ..., d_N$ as above but for $j \in \{2P + Q + 1, ..., N\}$ we find

$$\frac{\partial A_j}{\partial z_j}(z) + \sum_{\substack{i=1 \\ i \neq j}}^{N} \frac{d_i}{d_j} \left| \frac{\partial A_i}{\partial z_j}(z) \right| =$$

$$\left[-\frac{\partial H_j}{\partial z_j}(G(z)) + \sum_{\substack{i=1 \\ i \neq j}}^{N} \frac{d_i}{d_j} \left| \frac{\partial H_i}{\partial z_j}(G(z)) \right| \right] \Big/ \gamma_j'(g_j(z_j)) \leq \omega_j/p_j \qquad (2.12)$$

where

$$p_j = \begin{cases} \eta_j & \text{for } \omega \geq 0 \\ \xi_j & \text{for } \omega < 0 . \end{cases} \qquad (2.13)$$

If we take into account relation (2.1) together with inequalities (2.10) and (2.12) it follows $\mu_D(A'(z)) \leq \omega$ where

$$\omega = \max \left\{ \max_{j=1,\dots,2P+Q} \left[\max \left(\frac{s_j}{\tau_j} ; \frac{\omega_j}{c_j} \right) \right] ; \max_{j=2P+Q+1,\dots,N} \left(\frac{\omega_j}{p_j} \right) \right\} . \qquad (2.14)$$

(b) Let us consider P_i, $i = 1, 2, \dots, p$, the hyperplanes which define the CPWCD property of A. Let also x^0 and x^m be two different points in $\mathbf{R}^N$ not being on the same P_i (see Figure 2.4).

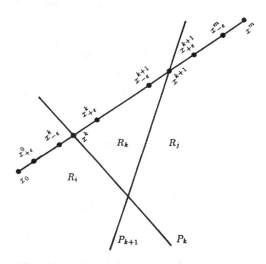

Figure 2.4 Sketch for the proof of Lemma 2.1

If $[x^0, x^m] = \{x \mid x = x(t) = x^0 + t(x^m - x^0) ; t \in [0,1]\}$ is the straight line segment connecting x^0 and x^m then $x(t_1)$, $x(t_2)$, $\dots, x(t_{m-1})$ are its interaction points with the P_i hyperplanes, where $0 = t_0 < t_1 < \dots < t_{m-1} < t_m = 1$. We take $0 < \varepsilon < \min_{k=0,\dots,m-1}(t_{k+1} - t_k)/2$ and for every $k = 0, \dots, m-1$ we denote: $x(t_k) = x^k$, $x(t_k + \varepsilon) = x^k_{+\varepsilon}$ and $x(t_k - \varepsilon) = x^k_{-\varepsilon}$. We can then write

$$x^k_{+\varepsilon} - x^{k+1}_{-\varepsilon} = (t_{k+1} - t_k - 2\varepsilon)(x^0 - x^m) \qquad (2.15)$$

and

$$A(x^0_{+\varepsilon}) - A(x^m_{-\varepsilon}) = \sum_{k=0}^{m-1} \left[A(x^k_{+\varepsilon}) - A(x^{k+1}_{-\varepsilon}) \right] + \sum_{k=1}^{m-1} \left[A(x^k_{-\varepsilon}) - A(x^k_{+\varepsilon}) \right] \ . \quad (2.16)$$

On the other hand, the fact that the points $x^k_{+\varepsilon}$ and $x^{k+1}_{-\varepsilon}$ are in the same (convex) region, and the CPWCD property of A, allow us to apply property (2.4):

$$A(x^k_{+\varepsilon}) - A(x^{k+1}_{-\varepsilon}) = \int_0^1 A'(x_\lambda)(x^k_{+\varepsilon} - x^{k+1}_{-\varepsilon}) \, d\lambda$$

where $x_\lambda = x^{k+1}_{-\varepsilon} + \lambda(x^k_{+\varepsilon} - x^{k+1}_{-\varepsilon})$. Hence, using (2.15) and (2.16) we derive

$$A(x^0_{+\varepsilon}) - A(x^m_{-\varepsilon}) = \sum_{k=0}^{m-1} (t_{k+1} - t_k - 2\varepsilon) \int_0^1 A'(x_\lambda)(x^0 - x^m) \, d\lambda$$

$$+ \sum_{k=1}^{m-1} \left[A(x^k_{-\varepsilon}) - A(x^k_{+\varepsilon}) \right] \ . \quad (2.17)$$

With this equality, by successively applying properties ii) and vi) from Lemma 1.1, and also (2.3), we can write

$$\langle x^0 - x^m, A(x^0_{+\varepsilon}) - A(x^m_{-\varepsilon}) \rangle_+ \leq \sum_{k=0}^{m-1} (t_{k+1} - t_k - 2\varepsilon) \times$$

$$\int_0^1 \langle x^0 - x^m, A'(x_\lambda)(x^0 - x^m) \rangle_+ \, d\lambda + \sum_{k=1}^{m-1} \| A(x^k_{-\varepsilon}) - A(x^k_{+\varepsilon}) \|_d \ .$$

Due to the property (2.2) and the first part of the present lemma, we also have

$$\langle x^0 - x^m, A(x^0_{+\varepsilon}) - A(x^m_{-\varepsilon}) \rangle_+$$

$$\leq \omega \| x^0 - x^m \|_d \sum_{k=0}^{m-1} (t_{t+1} - t_k - 2\varepsilon) + \sum_{k=1}^{m-1} \| A(x^k_{-\varepsilon}) - A(x^k_{+\varepsilon}) \|_d$$

$$= \omega(1 - 2\varepsilon m) \| x^0 - x^m \|_d + \sum_{k=1}^{m-1} \| A(x^k_{-\varepsilon}) - A(x^k_{+\varepsilon}) \|_d \ .$$

The continuity of the function A implies, via the property vii) from Lemma 1.1, the continuity of the function $\varepsilon \mapsto \langle x^0 - x^m, A(x^0_{+\varepsilon}) - A(x^m_{-\varepsilon}) \rangle_+$. This is why, for ε tending to zero, the last inequality becomes

$$\langle x^0 - x^m, A(x^0) - A(x^m) \rangle_+ \leq \omega \| x^0 - x^m \|_d \ ,$$

i.e. the total dissipativity of $A - \omega \mathcal{I}$.

If the points x^0 and x^m are on the same hyperplane, the continuity of A and the property vii) from Lemma 1.1 allow us to reduce this case to the above treated one.   □

We can also prove a result parallel to that from Lemma 2.1 but related to "greater than" inequalities.

Lemma 2.2. *If the assumptions I, II and V (or I*, II* and V*) are fulfilled, then*

(a) *there exists $\beta < 0$ such that for any region R_i of A and for any $z \in R_i$ (respectively, for any $z \in \mathbf{R}^N$) we have*

$$-\mu_D(-A'(z)) \geq \beta$$

(b) *for every x^0, $x^m \in \mathbf{R}^N$ it holds*

$$\langle x^0 - x^m, A(x^0) - A(x^m) \rangle_- \geq \beta \| x^0 - x^m \|_d \ .$$

Proof. (a) The desired inequality is found in a similar way as in the proof of Lemma 2.1 (a) and the constant β has the value

$$\beta = \min \left\{ \min_{j=1,\dots,2P+Q} \left[\min \left(\frac{r_j}{\tau_j} \ ; \ \frac{\beta_j}{c_j} \right) \right] \ ; \ \min_{j=2P+Q+1,\dots,N} \left(\frac{\beta_j}{\eta_j} \right) \right\} , \qquad (2.18)$$

where we have additionally denoted

$$r_j = -t_{jj} - \sum_{\substack{i=1 \\ i \neq j}}^{N} |t_{ij}| = \begin{cases} -1 - \dfrac{d_{j+1}}{d_j} \alpha_f^{(j+1)/2} & \text{for } j \in \{1,3,\dots,2P-1\} \\ -1 - \dfrac{d_{j-1}}{d_j} \alpha_r^{j/2} & \text{for } j \in \{2,4,\dots,2P\} \\ -1 & \text{for } j \in \{2P+1,\dots,2P+Q\} \ . \end{cases} \qquad (2.19)$$

(b) The proof begins with relation (2.17). If we successively apply properties v), vi), viii) from Lemma 1.1 and (2.3), we find

$$\langle x^0 - x^m, A(x^0_{+\varepsilon}) - A(x^m_{-\varepsilon}) \rangle_-$$
$$\geq \sum_{k=0}^{m-1} (t_{k+1} - t_k - 2\varepsilon) \int_0^1 \langle x^0 - x^m, A'(x_\lambda)(x^0 - x^m) \rangle_- \, d\lambda$$
$$- \sum_{k=1}^{m-1} \| A(x^k_{-\varepsilon}) - A(x^k_{+\varepsilon}) \|_d \ .$$

Property (2.2) and part (a) of the present lemma then yield

$$\langle x^0 - x^m, A(x^0_{+\varepsilon}) - A(x^m_{-\varepsilon}) \rangle_- \geq \beta(1 - 2\varepsilon m) \| x^0 - x^m \|_d - \sum_{k=1}^{m-1} \| A(x^k_{-\varepsilon}) - A(x^k_{+\varepsilon}) \|_d$$

which gives the result when ε tends to zero.   □

2.3. DC equations

The following result is related to the existence and uniqueness of the steady state solution of the network under consideration.

Theorem 2.1. *Under hypotheses* **I**, **II** *and* **IV** *(or* **I***, **II*** *and* **IV***) and for each $\overline{B} \in \mathbf{R}^N$, there exists a unique solution of the problem $(S(A, \overline{B}))$.*

Proof. Because **IV** implies **III**, Lemma 2.1 shows that $A - \omega\mathcal{I}$ is totally dissipative on $\mathbf{R}^N$. Moreover, this operator is continuous and then the result of Webb (see Webb [1972]) gives

$$\mathcal{R}(\mathcal{I} - \lambda(A - \omega\mathcal{I})) = \mathbf{R}^N \text{ for any } \lambda > 0 \ . \tag{2.20}$$

On the other hand, the inequalities (2.7) from the assumption **IV** imply $s_j < 0$ for any $j = 1, ..., 2P + Q$ (see (2.11)). Adding the fact that in assumption **IV**, ω_j is strictly negative for any j and therefore $p_j = \xi_j$ for $j = 2P + Q + 1, ..., N$ (see (2.13)), we have

$$\omega = \max\left\{\max_{j=1,...,2P+Q}\left[\max\left(\frac{s_j}{\tau_j} \ ; \ \frac{\omega_j}{c_j}\right)\right] \ ; \ \max_{j=2P+Q+1,...,N}\left(\frac{\omega_j}{\xi_j}\right)\right\} \tag{2.21}$$

which is a strictly negative number. Thus, in (2.20) we can take $\lambda = -1/\omega$ and we obtain the surjectivity of A, i.e. the existence part of the theorem. The uniqueness is an immediate consequence of the dissipativity of A and negativity of ω. Indeed, if z^1 and z^2 are two solutions of the steady state problem, then $0 = \langle z^1 - z^2, \overline{B} - \overline{B}\rangle_+ = \langle z^1 - z^2, A(z^1) - A(z^2)\rangle_+ \leq \omega\|z^1 - z^2\|_d$ that implies $z^1 = z^2$. $\square$

A first remark on the above theorem regards the fact that the existence and uniqueness of capacitor charges and inductor fluxes under steady state conditions proven above is equivalent (via the property of G being onto in $\mathbf{R}^N$) to the existence and uniqueness of the hybrid vector v which satisfies

$$T \cdot F(v) + H(v) = \overline{B} \ . \tag{2.22}$$

In the case when $R \neq 0$, $S \neq 0$ this equation describes the DC behaviour of a resistive N-port with constant independent sources contained in $\overline{B}$. If the first $2P + Q$ ports are connected to Ebers-Moll models of semiconductor devices, the following R pairs of terminals are open circuited and the last S ports are short circuited. It is convenient to include the last $R + S$ ports in the multiport, so that one studies a $2P + Q$-port connected with resistively modelled P transistors and Q diodes and which is described by H and $\overline{B}$ (we use the same notations as before).

This configuration corresponds to equation (2.22) where $R = S = 0$ and, since under such conditions T is invertible, we obtain the equivalent equation

$$F(v) + T^{-1} \cdot H(v) = T^{-1} \cdot \overline{B} . \tag{2.23}$$

The existence and uniqueness result formulated above for equations (2.22) and (2.23) is a partial extension of (or is strongly related to) many published results for the case when H is linear, for instance, Willson [1968, 1970], Sandberg and Willson [1969 a,b], Willson and Wu [1984]. In the case when H is nonlinear, our result is an extension of the smooth classes of functions, considered by Fujisawa and Kuh [1971], to CPWL and CPWCD classes for F and H respectively. Our result is of the same nature as that given in Chien [1977] (Corollaries 7 and 8), where F and H are CPWL.

Two properties that one might expect our network (described by (2.22) or (2.23)) to possess are: "small" changes in input $\overline{B}$ cause "small" changes in output v and, a bounded sequence of input vectors yields a bounded sequence of outputs. For the case when H is linear this problem is solved in Sandberg and Willson [1969 a] and in Willson [1970]. By using dissipativity we can, almost directly, derive the following result for the CPWCD case:

Theorem 2.2. (a) If the hypotheses **I**, **II** and **IV** (or **I***, **II*** and **IV***) are valid, then the solution v of (2.22) is a continuous function of the vector $\overline{B}$.
(b) If we add the hypotheses $f_j(0) = 0$ for $j = 1, ..., 2P + Q$ and $H(0) = 0$ (where **0** is the zero vector in $\mathbf{R}^N$) then every bounded sequence $\overline{B}^1, \overline{B}^2, \overline{B}^3, ...$ is mapped by equation (2.22) into a bounded output sequence $v^1, v^2, v^3, ...$

Proof. For two inputs $\overline{B}$ and $\overline{B}^*$ with outputs $v = G(z)$ and $v^* = G(z^*)$, we have

$$-\omega \|z - z^*\|_d \leq -\langle z - z^*, A(z) - A(z^*)\rangle_+ = -\langle z - z^*, -\overline{B} + \overline{B}^*\rangle_+ \leq \|\overline{B} - \overline{B}^*\|_d .$$

because of Lemma 2.1. Hence, the negativity of ω implies that $\overline{B} \mapsto z$ is a continuous function. By using the continuity of G, statement (a) follows. The additional hypotheses from (b) imply $A(0) = 0$ and taking $z^* = 0$ in above sequence of inequalities we derive

$$\|z\|_d \leq -\|\overline{B}\|_d / \omega . \tag{2.24}$$

But G is continuous on whole $\mathbf{R}^N$ and then v is bounded for bounded $\overline{B}$.

$\square$

2.4. Dynamic behaviour

Let us begin with an existence and uniqueness result:

Theorem 2.3. *Under hypotheses I, II and III (or I*, II* and III*) the problem* $(E(z^0, A, B))$ *has a unique solution* $z : [0, \infty[\mapsto \mathbf{R}^N,$ *for each* $z_0 \in \mathbf{R}^N.$

Proof. The hypotheses **I** and **II** assure the continuity of the function $(t, z) \mapsto A(z) + B(t)$ on the whole domain $[0, \infty[\times \mathbf{R}^N$. In addition, Lemma 2.1 gives, for each t, the dissipativity of $A(\cdot) + B(t) - \omega \mathcal{I}$ on $\mathbf{R}^N$. Hence, in view of Theorem 1.8, the result follows. $\square$

A comparison between the hypotheses used in the preceding theorem and those used in Sandberg [1969] allow us to formulate the following remarks:

(1) Here, the resistive multiport was supposed to be nonlinear and the condition $H(0) = 0$ (automatically satisfied in the linear case) was not imposed.

(2) We remark also that hypothesis **III** which restricts the class of nonlinearities is implicitly verified in the linear case considered by Sandberg.

(3) The nonlinear characteristic functions f_j, $j = 1, ..., 2P + Q$ and γ_j, $j = 2P + Q + 1, ..., N$, were supposed to be CPWL, while in Sandberg's paper they are of C^1 class.

(4) In the above we did not use the additional conditions $f_j(0) = 0$, $\gamma_j(0) = 0$ which were used by Sandberg.

(5) The additional boundedness hypothesis imposed by Sandberg to the time-function B (which describes the independent sources) is no longer needed here.

Let us now consider the solutions z^1 and z^2 of the problems $(E(z_0^1, A, B^1))$ and $(E(z_0^2, A, B^2))$ respectively. Of course,

$$\frac{d}{dt} \left[z^1(t) - z^2(t) \right] = A(z^1(t)) - A(z^2(t)) + B^1(t) - B^2(t)$$

and because the function $t \mapsto \|z^1(t) - z^2(t)\|_d$ is differentiable on $[0, \infty[$ except a countable set, Lemmas 1.15 and 1.1 give:

$$\frac{d}{dt}\|z^1(t) - z^2(t)\|_d \leq \langle z^1(t) - z^2(t), A(z^1(t)) - A(z^2(t))\rangle_+ \tag{2.25}$$
$$+ \|B^1(t) - B^2(t)\|_d \quad \text{a.e. in } [0, \infty[\, .$$

Analogously,

$$\frac{d}{dt}\|z^1(t) - z^2(t)\|_d \geq \langle z^1(t) - z^2(t), A(z^1(t)) - A(z^2(t))\rangle_- \tag{2.26}$$
$$- \|B^1(t) - B^2(t)\|_d \quad \text{a.e. in } [0, \infty[\, .$$

These inequalities will simply lead to the following qualitative results:

Theorem 2.4. *Let hypotheses* **I**, **II** *and* **IV** *(or* **I***, **II*** *and* **IV***) be fulfilled.*
*(a) If z^1 and z^2 are the solutions of the problem $(E(z_0^1, A, B^1))$ and $(E(z_0^2, A, B^2))$
respectively and $B^1(t) - B^2(t)$ tends to 0 for $t \to \infty$, then $\lim_{t\to\infty}[z^1(t) - z^2(t)] = 0$.*
*(b) If z is the solution of $(E(z_0, A, B))$ and $B_\infty = \lim_{t\to\infty} B(t)$, then there exists
$z_\infty \in \mathbf{R}^N$, independent of z_0, such that $z(t) \to z_\infty$ when $t \to \infty$.*
*(c) The solution of the problem $(E(z_0, A, B))$ is globally exponentially asympotically
stable in the Lyapunov' sense.*

Proof.
(a) Taking into account Lemma 2.1, we get from inequality (2.25):

$$\frac{d}{dt}\|z^1(t) - z^2(t)\|_d \leq \omega \|z^1(t) - z^2(t)\|_d + \|B^1(t) - B^2(t)\|_d \quad \text{a.e. in } [0, \infty[\quad (2.27)$$

which gives

$$\|z^1(t) - z^2(t)\|_d \leq \|z_0^1 - z_0^2\|_d \cdot e^{\omega t} + e^{\omega t} \int_0^t e^{-\omega s}\|B^1(s) - B^2(s)\|_d \, ds \qquad (2.28)$$

for all $t \in [0, \infty[$. Because of assumption **IV** we have $\omega < 0$ (see (2.21)) and the
result easily follows.
(b) Let z_∞ be the (unique) solution of stationary regime $A(z) + B_\infty = 0$ (see Theorem 2.1). Of course, z_∞ is identical with the solution of the problem $(E(z_\infty, A, B_\infty))$.
Then, the statement (a) gives the result.
(c) It is sufficient to put in (2.28) $B^1 = B^2$, to find that

$$\|z^1(t) - z^2(t)\|_d \leq \|z_0^1 - z_0^2\|_d \cdot e^{\omega t} \quad \text{for all } t > 0 , \qquad (2.29)$$

i.e. the stated stability property. $\square$

The above theorem particularly gives sufficient conditions under which a large
class of transistor circuits, under large signal operating conditions, possesses the
useful property of its output approaching a constant independent of the initial
condition when the input tends to a constant. This means, for instance, that one
cannot synthesize a bistable (with memory) network, if it satisfies the hypotheses
of the preceeding theorem.

The results of this theorem extend those obtained in Sandberg [1969] (Theorems
1, 1', 2', Corollaries 1, 1') in the sense of remarks (1)-(5) above. Moreover, our
assumption **IV** (or **IV***) is a natural extension of those considered by Sandberg for
a linear multiport. Other stability results can be found in Chua and Green [1976
a, Corollary 1], and other comments are given in Marinov [1990 b].

Further, let us give some sufficient conditions for the boundedness of the solution.

Theorem 2.5. *Under hypotheses* **I, II** *and* **IV** *(or* **I***, **II*** *and* **IV***) *and additionally supposing* $f_j(0) = 0$ *for* $j = 1, ..., 2P + Q$, $\gamma_j(0) = 0$ *for* $j = 2P + Q + 1, ..., N$ *and* $H(0) = 0$, *the boundedness of the function* $t \mapsto B(t)$ *on* $[0, \infty[$ *implies the same property for the solution function* $t \mapsto z(t)$ *of the problem* $(E(z_0, A, B))$.

Proof. Because the above conditions imply $A(0) = 0$, in the same way as that used in proving Theorem 2.4 (a) we can obtain

$$\frac{d}{dt}\|z\|_d \leq \omega\|z\|_d + \|B(t)\|_d \quad \text{a.e. in } [0, \infty[\ .$$

Then by using the boundedness of B, we find

$$\|z(t)\|_d \leq \|z(0)\|_d \cdot e^{\omega t} + e^{\omega t} \int_0^t \|B(\tau)\|e^{-\omega \tau} \, d\tau \ .$$

Because $\omega < 0$, the statement is clear.   □

Let us consider now the circuit under study where the independent sources are constant $B(t) = \overline{B}$. This corresponds to some step varying signals applied at the moment $t = 0$, while other sources remain constant.

Let us denote by $\bar{z}$ the solution of the problem $(E(\bar{z}_0, A, \overline{B}))$ and by $\bar{z}_\infty \neq \bar{z}_0$ the corresponding steady state solution –i.e. the solution of $(S(A, \overline{B}))$ equivalent with $(E(\bar{z}_\infty, A, \overline{B}))$. If relation (2.27) is valid, then

$$\frac{d}{dt}\|z(t) - z_\infty\|_d \leq \omega\|z_0 - z_\infty\|_d \quad \text{a.e. in } [0, \infty[\ . \tag{2.30}$$

If $\omega < 0$ (a sufficient condition for this is the hypothesis **IV**), this shows that the function $D : t \mapsto \|\bar{z}(t) - \bar{z}_\infty(t)\|_d / \|\bar{z}_0 - \bar{z}_\infty\|_d$ is strictly decreasing on $[0, \infty[$, starting from 1 and tending to 0. This function describes the global behaviour of the network between initial and steady state. If we fix $\lambda \in]0, 1[$ we can define "the λ-delay time" T_λ^d, as the (unique) moment when this function equals λ. The above results allow us to frame this parameter between two bounds which can be apriori computed.

Corollary 2.1. *Let hypotheses* **I, II, IV, V** *(or* **I***, **II***, **IV***, **V***) *be valid, where the latter two ones are satisfied for the same* $d = \{d_1, ..., d_N\}$. *Then,*

$$e^{\beta t} \leq D(t) \leq e^{\omega t} \quad \text{for all } t > 0 \ ,$$

that gives $\underline{T}_\lambda^d \leq T_\lambda^d \leq \overline{T}_\lambda^d$ *where* $\underline{T}_\lambda^d = \ln(\lambda/\beta)$ *and* $\overline{T}_\lambda^d = \ln(\lambda/\omega)$, β *and* ω *having the negative values from (2.18) and (2.21) respectively.*

Proof. The right hand inequalities immediatelly follow from (2.30), while for the left hand ones we use (2.26) and Lemma 2.2. Thus we derive

$$\frac{d}{dt}\|\bar{z}(t) - \bar{z}_\infty\|_d \geq \beta\|\bar{z}_0 - \bar{z}_\infty\|_d \quad \text{a.e. in } [0, \infty[\ ,$$

and the result is easily obtained.   □

As its definition shows, the delay time evaluates the global rate of evolution of the circuit between the initial and stationary states. As the bounds $\overline{T}^d_\lambda$ and $\underline{T}^d_\lambda$ depend simply on the circuit parameters, they are useful in estimating the switching speed of a circuit.

2.5. An example

In this section we provide an example which is application of the above theory. Let us consider the circuit from Figure 2.5, where the stage S_1 represents the preceding drive gates modelled by the step source e and a CPWL resistor whose characteristic is (coherent dimensions are used throughout the sequel):

$$i = \begin{cases} 50 \times 10^{-5} + 50 \times 10^{-5}u & \text{for } u < -2 \\ 25 \times 10^{-5}u & \text{for } u \in [-2, 2] \\ -50 \times 10^{-5} + 50 \times 10^{-5} & \text{for } u > 2 \ . \end{cases}$$

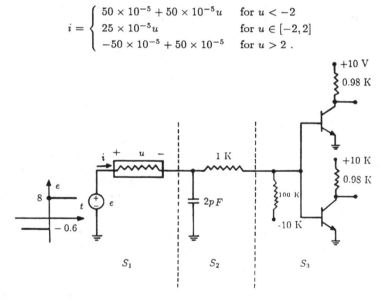

Figure 2.5

The following stage S_2 is a connecting resistor accompanied by a substrate capacitor. The stage S_3 is composed of two parallel inverters with identical parameters: $\alpha^1_f = 0.950$, $\alpha^1_r = 0.580$, $\tau_1 = 10^{-9}$, $\tau_1 = 10^{-8}$, $c_1 = 5 \times 10^{-12}$, $c_2 = 10^{-12}$,

$f_1(u_1) = -10^{-5} \times [\exp(-40u_1)-1]$, $f_2(u_2) = -1.638 \times 10^{-5} \times [\exp(-40u_2)-1]$. The bulk emitter, collector and base resistances are 10, 20 and 100 ohms respectively.

The operator H is CPWL in $\mathbf{R}^5$ with the hyperplanes $X_5 = 6$ and $X_5 = 10$ defining three regions, $k = 1, 2, 3$, for $X_5 < 6$, $X_5 \in [6, 10]$ and $X_5 > 10$ respectively. In each of these regions the resistive multiport is described by $w = -H^k v + B^k$, where H^k is a 5×5 symmetric matrix with the common elements for $k = 1, 2, 3$: $H_{11}^k = H_{33}^k = 5.68451 \times 10^{-3}$, $H_{22}^k = H_{44}^k = 0.99056 \times 10^{-3}$, $H_{12}^k = H_{21}^k = H_{34}^k = H_{43}^k = -0.94315 \times 10^{-3}$, $H_{13}^k = H_{31}^k = -4.22539 \times 10^{-3}$, $H_{14}^k = H_{41}^k = H_{23}^k = H_{32}^k = -0.04225 \times 10^{-3}$, $H_{15}^k = H_{51}^k = H_{35}^k = H_{53}^k = 0.47 \times 10^{-3}$, $H_{25}^k = H_{52}^k = H_{45}^k = H_{54}^k = 0.0047 \times 10^{-3}$. These matrices differ in $H_{55}^1 = H_{55}^3 = 1.44794 \times 10^{-3}$, $H_{55}^2 = 1.19794 \times 10^{-3}$. The elements of the vectors B^k are $B_1^k = B_3^k = -98.1 \times 10^{-4}$, $B_2^k = B_4^k = 99 \times 10^{-4}$ for $k = 1, 2, 3$ and $B_5^1 = 35.88 \times 10^{-4}$, $B_5^2 = 20.88 \times 10^{-4}$, $B_5^3 = 45.86 \times 10^{-4}$.

The circuit is initially in the steady state with $e = -0.6$ corresponding to the transistors in the off state: $v_1(0) = v_3(0) = 0.35$, $v_2(0) = v_4(0) = 10$ and $v_5(0) = -0.986$. The initial capacitor charges are $z_1(0) = z_3(0) = 1.7599 \times 10^{-12}$, $z_2(0) = z_4(0) = 10.1638 \times 10^{-12}$ and $z_5(0) = -1.972 \times 10^{-12}$.

Hypotheses **I** and **II** are obviously satisfied and the same is the case with assumption **IV** if we choose $d = \{0.961, 1, 0.961, 1, 0.823\}$ and $\omega_1 = \omega_3 = -31 \times 10^{-6}$, $\omega_2 = \omega_4 = -39 \times 10^{-6}$, $\omega_5 = -88 \times 10^{-6}$. With the same d, our circuit satisfies hypothesis **V**, namely $\beta_1 = \beta_3 = -11.34 \times 10^{-3}$, $\beta_2 = \beta_4 = -1.94 \times 10^{-3}$ and $\beta_5 = -2.31 \times 10^{-3}$. As we see, all hypotheses of Theorems 2.1–2.5 and Corollary 2.1 are fulfilled such that all properties given by these theorems are valid for our circuit. Also, $s_1 = s_3 = -0.0114$, $s_2 = s_4 = -0.442$, $r_1 = r_3 = -1.998$, $r_2 = r_4 = -1.557$ and $p_5 = \xi_5 = \eta_5 = 2 \times 10^{-12}$. With these, relations (2.18) and (2.21) yield: $\omega = -0.0114 \times 10^9$ and $\beta = -2.26 \times 10^9$ such that we can compute the bounds of the 0.1-delay time: $\underline{T}^d_{0.1} = 1.018 \times 10^{-9}$, $\overline{T}^d_{0.1} = 201.981 \times 10^{-9}$. On the other hand, an ad-hoc program numerically integrating the system $(E(z_0, A, B))$ gives $T^d_{0.1} = 146 \times 10^{-9}$. It seems that the reasonable tightness (especially of the upper bounds) and the calculational simplicity of the bounds from Corollary 2.1 make them useful for initial stages of circuit design. Such bounds can be included in so called "timing simulators", the usual handling programs for designers of digital circuits (see Ruehli and Ditlow [1983]).

Chapter III

ℓ^p-solutions of countable infinite systems of equations and applications to electrical circuits

3.0. Introduction

In the preceding chapter we have studied a lumped parameter model of a class of circuits containing a finite number of elements. Here we are interested in qualitative properties of the network in Figure 3.1.

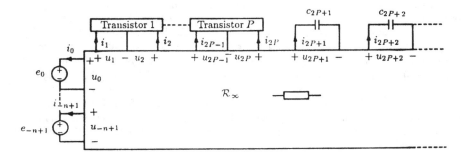

Figure 3.1 The infinite network under study

The circuit consists of n voltage sources, P bipolar transistors and an infinite number of capacitors and resistors, these last ones grouped together in an "infinite–port" $\mathcal{R}_\infty$. Each transistor $j = 1, 2, ..., P$ is a nonlinearly lumped parameter modelled by a Gummel circuit as shown in Figure 2.3 with positive parameters $\alpha_f^j < 1$, $\alpha_r^j < 1$, τ_{2j-1}, τ_{2j}, c_{2j-1}, c_{2j} and functions f_{2j-1}, f_{2j} with continuous and strict positive derivatives. We assume also that $\mathcal{R}_\infty$ is described by an infinite conductance matrix G whose elements G_{kj}, $k, j = -n + 1, -n + 2, ..., 0, 1, ...$ satisfy the following constraints:

$$\begin{cases} \sum_{k=1}^{\infty} |G_{kj}| < \infty \text{ for any } j \in \{-n+1, -n+2, ..., 0\} \cup \mathbb{N} \\ \sup \left\{ \sum_{k=1}^{\infty} |G_{kj}|; \ j \in \mathbb{N} \right\} < \infty \\ \inf \{G_{jj}; \ j \in \mathbb{N}\} > 0 \end{cases} \tag{3.1}$$

while for the capacitors we naturally impose

$$\inf \{c_j; \ j \in \mathbb{N}\} > 0. \tag{3.2}$$

According to (3.1), G is a linear function in ℓ^1 space, such that for any voltage sequence $\{u_j\}_{j=-n+1}^{\infty} \in \ell^1$ and for each $k = -n + 1, ..., 0, 1, ...$ we have

$$i_k = - \sum_{j=-n+1}^{\infty} G_{kj} u_j. \tag{3.3}$$

On the other hand, denoting by t_{kj} the elements of the matrix $T = \bigoplus_{j=1}^{P} \begin{bmatrix} 1 & -\alpha_r^j \\ -\alpha_f^j & 1 \end{bmatrix}$ for $k, j = 1, 2, ..., 2P$ and $t_{kj} = 0$ for $k, j > 2P$, the circuit structure implies:

$$i_k = \sum_{j=1}^{2P} t_{kj}(f_j \circ \gamma_j^{-1})(q_j) + \frac{dq_k}{dt} \qquad \text{for } k \in \mathbb{N}. \tag{3.4}$$

Here, q_j is the electric charge of capacitor c_j and $q_j = \gamma_j(u_j) = c_j u_j + \tau_j f_j(u_j)$ for $j = 1, 2, ..., 2P$ while $q_j = c_j u_j$ for $j > 2P$. Then, by using (3.2)–(3.4) we can derive the following infinite system of differential equations with variable $\{q_j\}_j \in \ell^1$ (finite total electric charge)

$$\begin{cases} \dfrac{dq_k(t)}{dt} = \sum_{j=1}^{\infty} a_{kj} q_j(t) + \tilde{f}_k(q_1, q_2, \dots) + \tilde{\tilde{f}}_k(t), \quad k = 1, 2, \dots \\ q_j(0) = \xi_j \end{cases} \tag{3.5}$$

where

$$
\left\{
\begin{array}{l}
a_{jj} = -G_{jj}/c_j \quad \text{for any } j \in \mathsf{N} \\[2mm]
a_{kj} = \begin{cases} 0 & \text{for any } k \in \mathsf{N} \text{ and } j \le 2P, \quad j \ne k \\ -G_{kj}/c_j & \text{for any } k \in \mathsf{N} \text{ and } j > 2P, \quad j \ne k \end{cases} \\[4mm]
\tilde{f}_k(q_1, q_2, \ldots) = \delta_k G_{kk} q_k / c_k - \sum_{j=1}^{2P} [t_{kj}(f_j \circ \gamma_j^{-1})(q_j) + G_{kj}\gamma_j^{-1}(q_j)] \\[2mm]
\qquad \text{for any } k \in \mathsf{N} \text{ and where } \delta_k = \begin{cases} 1 & \text{for } k \le 2P \\ 0 & \text{for } k > 2P \end{cases} \\[4mm]
\tilde{\tilde{f}}_k(t) = -\sum_{j=-n+1}^{0} G_{kj} e_j(t) \quad \text{for any } k \in \mathsf{N} \text{ and } t \ge 0 .
\end{array}
\right.
\tag{3.6}
$$

With this example in mind, we shall study below a system of the form (3.5) with less restrictive hypotheses than those imposed by (3.1). The mathematical interest of our problem consists of the dissipativity properties of the linear and nonlinear part of the system, as we show in Section 3.1. Section 3.2 includes the main result regarding uniqueness and stability of ℓ^p-continuous solutions of (S_ξ). For the quasiautonomous case an existence theorem is proved in Section 3.3 and it contains also a result about the stability of the equilibrium solution.

It is obvious that if the transistors do not exist in our circuit (i.e. $P = 0$), then system (3.5) is a linear one. In this case only, we are able to prove good convergence properties of the truncated system solution (Section 3.4). Also, easily computable bounds of truncation errors are infered. In Section 3.5 we verify that the system (3.5) (governing the circuit in Figure 3.1) fulfils the conditions imposed by the general theory developed in Sections 3.1–3.3. Consequently it has interesting qualitative properties. Also, in the linear case, a very simple concrete example illustrates the theory.

Infinite circuits with more general structures than the above one and whose variables belong to a Hilbert space (finite energy) were studied by Dolezal [1977,1979]. As in our case, the dissipativity is the central concept of the work. Also Zemanian [1976,1981,1982] treated two–dimensional infinite circuits appearing in the numerical analysis of certain boundary value problems in semiconductor devices. Moreover, many authors dealt with countable infinite systems of equations and their wide range of applications. The reference Deimling [1977] contains bibliography on this subject up to 1975. We also mention papers: McClure and Wong [1975,1976,1979], Chew, Shivakumar and Williams [1980], Miller and Michel [1980], Marinov [1984,1986].

3.1. Statement of the problem and preliminary results

Let $a_{ij}, (i,j = 1, 2, \dots)$ be complex valued functions defined on the interval $[0, \infty[, \xi = \{\xi_i\}_i \in \ell^p$ and $s \in [0, \infty[$. Let also $f_i, (i = 1, 2, \dots)$ be complex functions defined on $[0, \infty[\times \ell^p$. Our aim is to study the infinite system of equations, associated with an initial condition:

$$(S_\xi) \quad \begin{cases} \dfrac{du_i(t)}{dt} = \displaystyle\sum_{j=1}^{\infty} a_{ij}(t) u_j(t) + f_i(t, u_1(t), u_2(t), \dots) \\[3mm] u_i(s) = \xi_i \quad , \text{ where } i = 1, 2, \dots \text{ and } t \in [s, \infty[. \end{cases}$$

We call ℓ^p–solution of (S_ξ) a function $u : [s, \infty[\mapsto \ell^p$ with components u_i satisfying (S_ξ).

The system (S_ξ) is intricately associated with the abstract differential equation on ℓ^p:

$$(E_\xi) \quad \begin{cases} \dfrac{du(t)}{dt} = A(t)u(t) + F(t, u(t)) \\[3mm] u(s) = \xi \quad \text{and } t \in [s, \infty[\end{cases}$$

where $A(t)$ is a "matrix" linear function defined by using functions a_{ij} and F is a perturbation nonlinear function defined with the functions f_i. Unlike in the case of finite dimensional systems, equation (E_ξ) and system (S_ξ) are no longer equivalent. As a consequence, the proofs here extensively use properties of dissipative functions along with "classical" topics such as uniform convergence, derivability of series of functions, etc.

Let E be a normed vector space and we denote

$$C_E^p = \{f : E \mapsto \ell^p \,; \, f \text{ is continuous}\} \,.$$

The conditions of compactness of a set in ℓ^p (Lusternik, Sobolev [1974,p.167]) imply that $f \in C_E^p$ if its components $f_i : E \mapsto \mathbf{C}, i \in \mathbf{N}$ are continuous and $\sum_{i=1}^{\infty} |f_i(t)|^p < \infty$ uniformly on any compact subset of E (in short, u.c. E).

Let $p \in [1, \infty[$ and q be its conjugate, that is $q = p/(p-1)$ for $p > 1$ and $q = \infty$ for $p = 1$. Some assumptions, that are to be used in the following, are made regarding the functions a_{ij} and f_i from (S_ξ):

$\mathbf{A_1}$. $a_{ij} : [0, \infty[\mapsto \mathbf{C}$ is continuous for any $i, j \in \mathbf{N}$.

$\mathbf{A_2}$. If $p = 1$, $\displaystyle\sum_{\substack{i=1 \\ i \neq j}}^{\infty} |a_{ij}(t)| < \infty$ u.c. $[0, \infty[$, for any $j \in \mathbf{N}$ and

$$\beta(t) = \sup \left\{ \sum_{\substack{i=1 \\ i \neq j}}^{\infty} |a_{ij}(t)| \,; \, j \in \mathbf{N} \right\} \text{ is bounded on any compact}$$

subset of $[0, \infty[$.

If $p > 1$, $\displaystyle\sum_{\substack{j=1 \\ j\neq i}}^{\infty} |a_{ij}(t)|^q < \infty$ u.c. $[0,\infty[$ for any $i \in \mathbf{N}$ and

$$\beta(t) = \left\{ \sum_{i=1}^{\infty} \left[\sum_{\substack{j=1 \\ j\neq i}}^{\infty} |a_{ij}(t)|^q \right]^{p-1} \right\}^{1/p} < \infty \quad \text{u.c. } [0,\infty[.$$

$\mathbf{A_3}$. $\omega(t) = \sup\{\operatorname{Re} a_{jj}(t); \ j \in \mathbf{N}\}$ is bounded above on any compact subset of $[0,\infty[$.

$\mathbf{F_1}$. $f_i : [0,\infty[\times \ell^p \mapsto \mathbf{C}$ is continuous for each $i \in \mathbf{N}$ and

$$\sum_{i=1}^{\infty} |f_i(t,x)|^p < \infty \quad \text{u.c. } [0,\infty[\times \ell^p.$$

We will denote by $F : [0,\infty[\times \ell^p \mapsto \ell^p$ the function having f_i's as components; then, hypothesis $\mathbf{F_1}$ is equivalent to $F \in C^p_{[0,\infty[\times \ell^p}$.

$\mathbf{F_2}$. There is a function $\alpha : [0,\infty[\to \mathbf{R}$ that is integrable on any compact interval from $[0,\infty[$ such that $F(t,\cdot) - \alpha(t)\mathcal{I}$ is dissipative for any $t \in [0,\infty[$.

The next result can be easily shown (for $p = 1$ see Taylor [1958,p.220] and McClure, Wong [1976]).

Lemma 3.1. *Let hypotheses* $\mathbf{A_1}$ *and* $\mathbf{A_2}$ *be valid; then,*

(i) for any $t \geq 0$ *we can define the linear bounded operator*

$$B(t) : \ell^p \mapsto \ell^p \quad \text{with} \quad B(t)x = \left\{ \sum_{\substack{j=1 \\ j\neq i}}^{\infty} a_{ij}(t)x_j \right\}_i$$

where $x = \{x_j\}_j \in \ell^p$. *Morover,* $\|B(t)\| \leq \beta(t)$ *with equality for* $p = 1$.

(ii) for any $s \in [0,\infty[$ *and* $u \in C^p_{[s,\infty[}$, *the function* $t \mapsto B(t)u(t)$ *belongs to* $C^p_{[s,\infty[}$. $\qquad\square$

Let us consider now the function $\mu_A : [0,\infty[\mapsto [-\infty,\infty]$ defined by

$$\mu_A(t) = \begin{cases} \sup\left\{ \operatorname{Re} a_{jj}(t) + \displaystyle\sum_{\substack{i=1 \\ i\neq j}}^{\infty} |a_{ij}(t)| \ ; \ j \in \mathbf{N} \right\} & \text{for } p = 1 \\[4mm] \sup\left\{ \operatorname{Re} a_{jj}(t) \ ; \ j \in \mathbf{N} \right\} + \beta(t) & \text{for } p > 1 \end{cases}$$

where $\beta(t)$ is obviously defined by hypotheses $\mathbf{A_2}$ in the case $p > 1$.

Lemma 3.2. *Under hypotheses* $\mathbf{A_1}, \mathbf{A_2}, \mathbf{A_3}$, μ_A *is Riemann integrable on any compact interval of* $[0, \infty[$.

Proof. Taking into account hypotheses $\mathbf{A_1}$ and $\mathbf{A_2}$, it follows that μ_A, ω and β are lower semicontinuous and consequently almost everywhere continuous in $[0, \infty[$ and bounded below on any compact set from $[0, \infty[$. As $\mu_A(t) \leq \omega(t) + \beta(t)$ for any $t \in [0, \infty[$ it follows from hypothesis $\mathbf{A_3}$ that μ_A is also bounded above on compacts of $[0, \infty[$. The proof is complete. $\square$

Let us now consider the set:

$$\mathcal{D} = \left\{ x = \{x_i\}_i \in \ell^p \quad ; \quad \{a_{ii}(t)x_i\}_i \in \ell^p \quad \text{for each } t \in [0, \infty[\right\} .$$

We note that $\mathcal{D}$ is a linear subspace of ℓ^p and $\overline{\mathcal{D}} = \ell^p$ in the topology of the norm in ℓ^p. From hypothesis $\mathbf{A_2}$ it follows immediately that

$$\mathcal{D} = \left\{ x = \{x_i\}_i \in \ell^p \quad ; \quad \left\{ \sum_{j=1}^{\infty} a_{ij}(t)x_j \right\}_i \in \ell^p \quad \text{for each } t \in [0, \infty[\right\}$$

and it results (under hypothesis $\mathbf{A_2}$) that for each $t \in [0, \infty[$ we may consider the linear functions:

$$D(t) : \mathcal{D} \mapsto \ell^p \quad \text{with} \quad D(t)x = \{a_{ii}(t)x_i\}_i \quad \text{and}$$

$$A(t) : \mathcal{D} \mapsto \ell^p \quad \text{with} \quad A(t)x = \left\{ \sum_{j=1}^{\infty} a_{ij}(t)x_j \right\}_i ,$$

where $x = \{x_i\}_i \in \mathcal{D}$. We remark that $A(t) = D(t) + B(t)$ where $B(t)$ satisfies Lemma 3.1.

If $t \in [0, \infty[$ and $n \in \mathbb{N}$ then we may start from the operator $A(t)$ and, with assumption $\mathbf{A_2}$, construct the function $A^n(t) : \ell^p \mapsto \ell^p$ with the ith component defined by:

$$[A^n(t)x]_i = \begin{cases} \displaystyle\sum_{j=1}^{\infty} a_{ij}(t)x_j & \text{if } i \leq n \\ 0 & \text{if } i > n \end{cases}$$

where $x = \{x_i\}_i \in \ell^p$. This definition is consistent because, as it follows from hypothesis $\mathbf{A_2}$ that

$$\sum_{j=1}^{\infty} a_{ij}(t)x_j \quad \text{is convergent for any} \quad x = \{x_j\}_j \in \ell^p, \ i \in \mathbb{N}, \ t \in [0, \infty[.$$

Analogously we introduce the everywhere defined functions $D^n(t), B^n(t), F^n(t)$: $\ell^p \mapsto \ell^p$, their components being defined as

$$[D^n(t)x]_i = \begin{cases} a_{ii}(t)x_i & \text{if } i \leq n \\ 0 & \text{if } i > n \end{cases}$$

$$[B^n(t)x]_i = [A^n(t)x]_i - [D^n(t)x]_i$$

$$[F^n(t)x]_i = \begin{cases} f_i(t,x) & \text{for } i \leq n \\ 0 & \text{for } i > n \end{cases}$$

for every $x = \{x_i\}_i \in \ell^p$.

We shall associate with the sequence $\{x_i\}_i \in \ell^p$, the sequence $\overline{x}^n = \{\overline{x}_i^n\}_i$ in which $\overline{x}_i^n = x_i$ for $i \leq n$ and $\overline{x}_i^n = 0$ for $i > n$.

Lemma 3.3. *Under hypothesis* $\mathbf{A_2}$ *we have for* $x = \{x_i\}_i \in \ell^p$, $t \in [0, \infty[$ *and* $n \in \mathbf{N}$:

$$\langle \overline{x}^n, A^n(t)x \rangle_+ \leq \mu_A(t)\|\overline{x}^n\| + \beta(t)\|x - \overline{x}^n\| \ .$$

Proof.

Let us first consider $p = 1$. Let $Y = \{1, 2, \dots\}$ with counting measure and $Y_0 = \{i \in Y : x_i = 0\}$. From Lemma 1.7 we obtain:

$$\langle \overline{x}^n, A^n(t)x \rangle_+ = \sum_{i \in Y_0} |[A^n(t)x]_i| + \sum_{i \in Y - Y_0} |x_i| \operatorname{Re}[A^n(t)x]_i / x_i \leq$$

$$\leq \sum_{i \in Y} |x_i| \operatorname{Re} a_{ii}(t) + \sum_{i \in Y} |[B^n(t)x]_i| \leq$$

$$\leq \sum_{j=1}^n |x_j| \left(\operatorname{Re} a_{jj}(t) + \sum_{\substack{i=1 \\ i \neq j}}^n |a_{ij}(t)| \right) + \sum_{j=n+1}^\infty |x_j| \sum_{\substack{i=1 \\ i \neq j}}^n |a_{ij}(t)| \leq$$

$$\leq \mu_A(t)\|\overline{x}^n\| + \beta(t)\|x - \overline{x}^n\| \ .$$

Now let $p > 1$. If $\overline{x}^n = 0$ then

$$\langle \overline{x}^n, A^n(t)x \rangle_+ = \|A^n(t)x\| = \|B^n(t)x\| \leq \beta(t)\|x\|$$

and if $\overline{x}^n \neq 0$ let Y and Y_0 be defined as above. Following Lemma 1.8 we have:

$$\langle \overline{x}^n, D^n(t)x \rangle_+ = \frac{1}{\|\overline{x}^n\|^{p-1}} \sum_{i \in Y - Y_0} |x_i|^p \operatorname{Re} a_{ii}(t)x_i / x_i \leq \omega(t)\|\overline{x}^n\|$$

and this results in

$$\langle \overline{x}^n, A^n(t)x \rangle_+ \leq \langle \overline{x}^n, D^n(t)x \rangle_+ + \langle \overline{x}^n, B^n(t)x \rangle_+ \leq$$
$$\leq \omega(t)\|\overline{x}^n\| + \|B^n(t)x\| \leq$$
$$\leq \omega(t)\|\overline{x}^n\| + \beta(t)\|x\| \leq$$
$$\leq [\omega(t) + \beta(t)]\|\overline{x}^n\| + \beta(t)\|x - \overline{x}^n\| .$$

□

A similar procedure may be used to prove the next result.

Lemma 3.4. *Under hypothesis* A_2 *we have for* $x = \{x_i\}_i \in \mathcal{D}$, $t \geq 0$:
(i) $\langle x, D(t)x \rangle_+ \leq \omega(t)\|x\|$ *i.e.* $D(t) - \omega(t)\mathcal{I}$ *is totally dissipative.*
(ii) $\langle x, A(t)x \rangle_+ \leq \mu_A(t)\|x\|$ *i.e.* $A(t) - \mu_A(t)\mathcal{I}$ *is totally dissipative.*

3.2. Properties of continuous ℓ^p–solutions

It is the purpose of this section to investigate the behaviour of the solutions of (S_ξ). The following theorem is our main result:

Theorem 3.1. *Let* $\xi, \eta \in \ell^p$ *and* $u, v \in C^p_{[s,\infty[}$ *be* ℓ^p*–solutions for* (S_ξ) *and* (S_η) *respectively. Then, under hypotheses* A_1, A_2, A_3, F_1, F_2,

$$\|u(t) - v(t)\| \leq \|\xi - \eta\| \exp \int_s^t [\mu_A(r) + \alpha(r)] \, dr$$

for any $t \in [s, \infty[$.

The previous theorem results in the uniqueness of solutions as well as some stability criteria, dealt with in the following assertion:

Corollary 3.1. *Under hypotheses* A_1, A_2, A_3, F_1, F_2, *let* $u \in C^p_{[s,\infty[}$ *be a* ℓ^p*–solution for* (S_ξ), *where* $\xi \in \ell^p$. *Then,*
(i) u *is unique.*
(ii) *if* $\displaystyle\limsup_{t \to \infty} \int_s^t [\mu_A(r) + \alpha(r)] \, dr < \infty$, *then* u *is stable.*
(iii) *if* $\displaystyle\lim_{t \to \infty} \int_s^t [\mu_A(r) + \alpha(r)] \, dr = -\infty$, *then* u *is asymptotically stable.*
(iv) *if* $[\mu_A(r) + \alpha(r)] \leq 0$ *for any* $r \geq s$, *then* u *is uniformly stable.*

(v) *if $\delta < 0$ exists such that $\mu_A(r) + \alpha(r) \leq \delta$ for any $r \geq s$, then u is uniformly asymptotically stable.*

Proof of Theorem 3.1.

If u and v are solutions for (S_ξ) and (S_η) respectively, then for any $n \in \mathbf{N}$ and for any $t \in [s, \infty[$,

$$\frac{d}{dt}\overline{u(t) - v(t)}^n = A^n(t)[u(t) - v(t)] + F^n(t, u(t)) - F^n(t, v(t)) . \tag{3.7}$$

The function $t \mapsto \|\overline{u(t) - v(t)}^n\|$ is derivable on $[s, \infty[\backslash\mathcal{N}_n$ where $\mathcal{N}_n$ is a, at most, countable set. Moreover, it is derivable on $[s, \infty[\backslash\mathcal{N}$, where $\mathcal{N} = \bigcup_{n=1}^\infty \mathcal{N}_n$. Following Lemma 1.15 we derive from (3.7):

$$\frac{d}{dt}\|\overline{w(t)}^n\| = \left\langle \overline{w(t)}^n, A^n(t)w(t) + F^n(t, u(t)) - F^n(t, v(t)) \right\rangle_- \tag{3.8}$$

for $t \in [s, \infty[\backslash\mathcal{N}$, where we have denoted $w(t) = u(t) - v(t)$. From (3.8) and Lemma 1.1 it follows:

$$\frac{d}{dt}\|\overline{w(t)}^n\| \leq \left\langle \overline{w(t)}^n, A^n(t)w(t) \right\rangle_+ + \left\langle \overline{w(t)}^n, F(t, \overline{u(t)}^n) - F(t, \overline{v(t)}^n) \right\rangle_- + \delta_n(t) \tag{3.9}$$

where we have denoted

$$\delta_n(t) = \|F^n(t, u(t)) - F(t, u(t))\| + \|F(t, u(t)) - F(t, \overline{u(t)}^n)\| +$$
$$+ \|F^n(t, v(t)) - F(t, v(t))\| + \|F(t, v(t)) - F(t, \overline{v(t)}^n)\| .$$

Taking into consideration Lemma 3.3, hypothesis $\mathbf{F_2}$, Lemma 1.4 and denoting $\gamma_n(t) = \delta_n(t) + \beta(t)\|w((t) - \overline{w(t)}^n\|$, we obtain from (3.9)

$$\frac{d}{dt}\|\overline{w(t)}^n\| \leq [\mu_A(t) + \alpha(t)] \|\overline{w(t)}^n\| + \gamma_n(t) \quad \text{on } [s, \infty[\backslash\mathcal{N}. \tag{3.10}$$

The function $t \mapsto \int_t^{t_1}[\mu_A(r) + \alpha(r)]\, dr$ is absolutely continuous on every interval $[t_1, t_2] \subset [s, \infty[\backslash\mathcal{N}$. Consequently, it is derivable almost everywhere in $[t_1, t_2]$ and from (3.10) one obtains:

$$m_n(t) = \frac{d}{dt}\|\overline{w(t)}^n\| \exp \int_t^{t_1}[\mu_A(r) + \alpha(r)]\, dr \leq$$
$$\leq \gamma_n(t) \exp \int_t^{t_1}[\mu_A(r) + \alpha(r)]\, dr \quad \text{a.e. on } [t_1, t_2] . \tag{3.11}$$

On the other hand, from (3.8) and Lemma 1.1 we derive:

$$\left|\frac{d}{dt}\|\overline{w(t)}^n\|\right| \leq \|A^n(t)w(t)\| + \|F^n(t, u(t))\| + \|F^n(t, v(t))\| \qquad (3.12)$$

for $t \in [s, \infty[\setminus\mathcal{N}$.

But the functions $t \mapsto A^n(t)u(t)$, $t \mapsto F^n(t, u(t))$ and $t \mapsto F^n(t, v(t))$ are continuous, such that (3.12) clearly implies the Lipschitz property of the function $t \mapsto \|\overline{w(t)}^n\|$ on $[t_1, t_2]$. This shows that the function

$$t \mapsto \|\overline{w(t)}^n\| \exp \int_t^{t_1} [\mu_A(r) + \alpha(r)] \, dr$$

is absolute continuous on $[t_1, t_2]$. Therefore

$$\|\overline{w(t_2)}^n\| \exp \int_{t_2}^{t_1} [\mu_A(r) + \alpha(r)] \, dr - \|\overline{w(t_1)}^n\| = \int_{t_1}^{t_2} m_n(r) \, dr . \qquad (3.13)$$

According to the continuity of u we have

$$\left\{\overline{u(t)}^n\right\}_n \to u(t) \quad \text{u.c.} \quad [s, \infty[\qquad (3.14)$$

and analogously for v. From hypothesis $\mathbf{F_1}$ we may conclude:

$$\{F^n(t, u(t))\}_n \to F(t, u(t)) \quad \text{u.c.} \quad [s, \infty[\qquad (3.15)$$

and similarly with v instead of u.

Finally, a result of McClure and Wong,[1979,Lemma 2], implies:

$$\left\{F(t, \overline{u(t)}^n)\right\}_n \to F(t, u(t)) \quad \text{u.c.} \quad [s, \infty[\qquad (3.16)$$

and the same with v instead of u. It follows from (3.14)-(3.16) that $\{\gamma_n(t)\}_n \to 0$ u.c. $[s, \infty[$. Hence, from (3.13) and (3.11) we have:

$$\|w(t_2)\| \exp \int_{t_2}^{t_1} [\mu_A(r) + \alpha(r)] \, dr - \|w(t_1)\| \leq 0.$$

Following continuity arguments, this inequality may be extended on every interval $[s, t]$ of $[s, \infty[$. Thus, the proof is complete.

$\square$

3.3. Existence of continuous ℓ^p–solutions for the quasi-autonomous case

The existence of continuous ℓ^p–solutions of the infinite system of equations

$$(S_\xi^*) \qquad \begin{cases} \dfrac{du_i(t)}{dt} = \displaystyle\sum_{j=1}^{\infty} a_{ij}u_j(t) + \tilde{f}_i(u_1(t), u_2(t), \dots) + \tilde{\tilde{f}}_i(t) \\[2mm] u_i(s) = \xi_i \quad ; \quad i \in \mathsf{N}, t \in [s, \infty[, \xi = \{\xi_i\}_i \in \ell^p \ , \end{cases}$$

where $a_{ij} \in \mathsf{C}$, $\tilde{f}_i : \ell^p \mapsto \mathsf{C}$ and $\tilde{\tilde{f}}_i : [s, \infty] \mapsto \mathsf{C}$ for any $i, j \in \mathsf{N}$, will be investigated below. This is a particular case (called "quasiautonomous") of the system (S_ξ).

We assume that the following hypotheses are satisfied:

A$_1$*. If $p = 1$ $\displaystyle\sum_{\substack{i=1 \\ i \neq j}}^{\infty} |a_{ij}| < \infty$ for any $j \in \mathsf{N}$ and

$$\beta = \sup\left\{ \sum_{\substack{i=1 \\ i \neq j}}^{\infty} |a_{ij}| \quad ; \quad j \in \mathsf{N} \right\} < \infty.$$

If $p > 1$ $\displaystyle\sum_{\substack{j=1 \\ j \neq i}}^{\infty} |a_{ij}|^q < \infty$ for any $i \in \mathsf{N}$ and

$$\beta = \left\{ \sum_{i=1}^{\infty} \left[\sum_{\substack{j=1 \\ j \neq i}}^{\infty} |a_{ij}|^q \right]^{p-1} \right\}^{1/p} < \infty.$$

A$_2$*. $\omega = \sup\{\operatorname{Re} a_{ii} \ ; \ i \in \mathsf{N}\} < 0$

F$_1$*. The function $\tilde{f}_i : \ell^p \mapsto \mathsf{C}$ is continuous for any $i \in \mathsf{N}$ and

$$\sum_{i=1}^{\infty} |\tilde{f}_i(x)|^p < \infty \quad \text{u.c.} \ \ell^p.$$

We shall denote by $\tilde{F} : \ell^p \mapsto \ell^p$ the function with $\tilde{f}_i$'s as components; hypothesis **F$_1$*** states that $\tilde{F} \in C_{\ell^p}^p$.

F$_2$*. There is a $\alpha \in \mathsf{R}$ such that $\tilde{F} - \alpha \mathcal{I}$ is totally dissipative.

F$_3^*$. $\displaystyle\sum_{i=1}^{\infty} |\tilde{\tilde{f}}_i(t)|^p < \infty$ for every $t \in [s,\infty[$. We shall denote by
$\tilde{\tilde{F}} : [s,\infty[\mapsto \ell^p$ the function with $\tilde{\tilde{f}}_i$'s as components.

F$_4^*$. There is a continuous function $d : [s,\infty[\times[s,\infty[\mapsto [s,\infty[$ such that

$$\|\tilde{\tilde{F}}(t_1) - \tilde{\tilde{F}}(t_2)\| \le |t_1 - t_2| d(t_1, t_2) \quad \text{for all } (t_1, t_2) \in [s,\infty[\times[s,\infty[.$$

Let us denote by $F : [s,\infty[\times\ell^p \mapsto \ell^p$ the function defined by $F(t,x) = \tilde{F}(x) + \tilde{\tilde{F}}(t)$, with components $f_i = \tilde{f}_i + \tilde{\tilde{f}}_i$. We will denote

$$\mathcal{D} = \left\{ x = \{x_i\}_i \in \ell^p \;\; ; \;\; \{a_{ii}x_i\}_i \in \ell^p \right\}$$

and B, A, D will be functions defined as $B(t), A(t), D(t)$ respectively, in the non-autonomous case. The real μ_A is defined analogously. With proper modifications Lemmas 3.1 and 3.4 are valid also for the newly defined functions.

Lemma 3.5. *Under assumptions* $\mathbf{A}_1^*, \mathbf{A}_2^*, \mathbf{F}_1^*, \mathbf{F}_2^*$ *and with* $c = \omega + \beta + \alpha$ *the operator* $\Lambda = A + F(t,.) - c\mathcal{I}$ *is "m"-totally dissipative, for all* $t \in [s,\infty[$.

Proof. Let us first suppose that the following properties are already proved

(i) $\Lambda_1 = D - \omega\mathcal{I}$ is linear, closed, densely defined and "m"-totally dissipative.

(ii) $\Lambda_2 = B + F(t,\cdot) - (\beta + \alpha)\mathcal{I}$ is continuous, everywhere defined and totally dissipative for each $t \in [s,\infty[$.

Then, following a result of Webb [1972], we have "m"-total dissipativity of $\Lambda = \Lambda_1 + \Lambda_2$.

Most of the properties assumed, as satisfied in (i) and (ii), follow directly from the definition of the set $\mathcal{D}$, functions D, B, F, as well as from Lemma 3.1. We have to verify only the underlined properties in (i) and (ii). Let $\{x^n\}_n$ be a sequence of elements of $\mathcal{D}$ such that $\{x^n\}_n \to x$ and $\{\Lambda_1 x^n\}_n \to y$. There is then $\gamma > 0$ with $\|Dx^n\|^p < \gamma$ for any $n \in \mathbf{N}$. On the other hand, for any $m \in \mathbf{N}$ we have

$$\sum_{i=1}^{m} |a_{ii}x_i|^p \le 2^p \sum_{i=1}^{m} |a_{ii}(x_i - x_i^n)|^p + 2^p \sum_{i=1}^{m} |a_{ii}x_i^n|^p$$
$$\le 2^p\|x - x^n\|^p \max\{|a_{ii}|^p \; ; \; i = 1, 2, \ldots, m\} + 2^p\|Dx^n\|^p$$
$$\le 2^p\gamma$$

for $n \to \infty$. It results in $\sum_{i=1}^{\infty} |a_{ii}x_i|^p < \infty$ and therefore $x \in \mathcal{D}$. But

$$\{a_{ii}x_i^n - \omega x_i^n\}_n \to a_{ii}x_i - \omega x_i \quad \text{for any } i \in \mathbf{N},$$

i.e. $y = \Lambda_1 x$. That is, Λ_1 is closed. The total dissipativity of Λ_1 is obtained in Lemma 3.4. In order to prove the "m"-property we have to verify that for a given $y = \{y_i\}_i \in \ell^p$ there is $\rho > 0$ and $x = \{x_i\}_i \in \mathcal{D}$ such that $(1 - \rho a_{ii} + \rho \omega)x_i = y_i$ for any $i \in \mathbb{N}$. A proper choice is $\rho = -1/\omega$ and $x_i = \omega y_i/a_{ii}$, because $|a_{ii}| \geq -\omega > 0$ and $|a_{ii}x_i|^p = |\omega|^p |y_i|^p$ for any $i \in \mathbb{N}$, that is $x = \{x_i\}_i \in \mathcal{D}$.

The total dissipativity of Λ_2 follows from

$$\langle x - y, Bx + F(t,x) - By - F(t,y)\rangle_+ \leq \|B(x-y)\| + \langle x - y, F(t,x) - F(t,y)\rangle_+$$
$$= \|B(x-y)\| + \langle x - y, \tilde{F}(x) - \tilde{F}(y)\rangle_+ \leq (\beta + \alpha)\|x - y\|$$

where Lemma 1.1, Lemma 3.1 and assumptions $\mathbf{F_2^*}$ were used.   □

Lemma 3.6. *Let hypotheses $\mathbf{A_1^*}, \mathbf{A_2^*}, \mathbf{F_1^*}, \mathbf{F_2^*}$ be valid and $c = \omega + \beta + \alpha < 0$. Then for any $y = \{y_i\}_i \in \ell^p$ there is $x \in \mathcal{D}$ such that its components verify the system*

$$\sum_{j=1}^{\infty} a_{ij}x_i + \tilde{f}_i(x) = y_i \quad , \quad i \in \mathbb{N} \; .$$

Proof. It is sufficient to show that the equation $Ax + \tilde{F}(x) = y$ has a solution. By the same arguments as those used in the proof of Lemma 3.5 we can show that $A + \tilde{F} - c\mathcal{I}$ is "m"-totally dissipative and this implies the existence of the solution of the equation $x - \rho[Ax + \tilde{F}(x) - cx] = z$ for any $z \in \ell^p$ and $\rho > 0$. Taking $\rho = -1/c$ and $z = y/c$ we have the proof.   □

Lemma 3.7. *Let hypotheses $\mathbf{A_1^*}$ and $\mathbf{F_1^*}$ be valid. If $t \in [s, \infty[$ and $\{x^n\}_n$ is a sequence in $\mathcal{D}$ such that $\{x^n\}_n \to x$ and $\|Ax^n + F(t, x^n)\| < M$ for any $n \in \mathbb{N}$, then $x \in \mathcal{D}$ and $\{Ax^n + F(t, x^n)\}_n \overset{w}{\to} Ax + F(t, x)$.*

Proof. As the function $B + F(t, \cdot)$ is continuous on ℓ^p it is sufficient to verify the above property for D instead of $A + F(t, \cdot)$. But $x \in \mathcal{D}$ as we have obtained in the proof of Lemma 3.5. As $\{a_{ii}x_i^n\}_n \to a_{ii}x_i$ for any $i \in \mathbb{N}$ and $\|Dx\| < M$ it follows (cf. Lusternik, Sobolev [1974,pp.149]) that $\{Dx^n\}_n \overset{w}{\to} Dx$.   □

Now we may derive the main result of this section.

Theorem 3.2. *Under assumptions $\mathbf{A_1^*}, \mathbf{A_2^*}, \mathbf{F_1^*} - \mathbf{F_4^*}$ and for any $\xi \in \ell^p$, there exists $u \in C_{[s,\infty[}^p$, ℓ^p-solution of S_ξ^*.*

Proof. Let $\{\xi^n\}_n$ be a sequence of $\mathcal{D}$ such that $\{\xi^n\}_n \to \xi$. Because of Lemma 3.5, Lemma 3.7 and assumptions $\mathbf{F_3^*}, \mathbf{F_4^*}$, the hypotheses $\mathbf{H1, H2}$ and $\mathbf{H3}$ of Theorem

1.9 are satisfied and this ensures the existence of a function $u^n : [s, \infty[\mapsto \mathcal{D}$ with the following properties:

a) $u^n \in C^p_{[s,\infty[}$.

b) u^n is weakly derivable (let $(u^n)'_w$ be its weak derivative), $u^n(s) = \xi^n$ and $(u^n)'_w(t) = Au^n(t) + F(t, u^n(t))$ for any $t \in [s, \infty[$.

c) if u^m also satisfies a) and b) with $u^m(s) = \xi^m$ then,

$$\|u^n(t) - u^m(t)\| \le \|\xi^n - \xi^m\| \exp c(t - s) \quad \text{for any} \quad t \in [s, \infty[,$$

where $c = \omega + \beta + \alpha$.

From b) we have

$$\frac{d}{dt} f^*[u^n(t)] = f^*[Au^n(t) + F(t, u^n(t))] \tag{3.17}$$

for any $f^* \in \ell^{p^*}$ (the dual space of ℓ^p), with the unique associated sequence $\{m_k\}_k \in \ell^q$. Taking $m_k = \delta^i_k$, where $i \in \mathbb{N}$ and δ^i_k is the Kronecker's symbol, we have from (3.17)

$$\frac{d}{dt} u^n_i(t) = a_{ii} u^n_i(t) + [Bu^n(t)]_i + f_i(t, u^n(t)) \tag{3.18}$$

for any $t \ge s$ and $i \in \mathbb{N}$. From c) we observe that $\{u^n(t)\}_n$ is a uniform Cauchy sequence on any compact set of $[s, \infty[$. There is then, $u : [s, \infty[\mapsto \overline{\mathcal{D}} = \ell^p$ such that $\{u^n(t)\}_n \to u(t)$ u.c. $[s, \infty[$. We have also that $u \in C^p_{[s,\infty[}$ and $u(s) = \xi$. On the other hand, B is a continuous function so that $\{[Bu^n(t)]_i\}_n \to [Bu(t)]_i$ u.c. $[s, \infty[$, for each $i \in \mathbb{N}$. From $|f_i(t, u^n(t)) - f_i(t, u(t))| \le \|\tilde{F}(u^n(t)) - \tilde{F}(u(t))\|$ and from McClure, Wong [1979,Lemma 2], we conclude that

$$\{f_i(t, u^n(t))\}_n \to f_i(t, u(t)) \quad \text{u.c.} \quad [s, \infty[$$

for each $i \in \mathbb{N}$. Then, from (3.18) we have

$$\left\{ \frac{du^n_i(t)}{dt} \right\}_n \to \frac{du_i(t)}{dt} \quad \text{u.c.} \quad [s, \infty[$$

and therefore

$$\frac{du_i(t)}{dt} = a_{ii} u_i(t) + [Bu(t)]_i + f_i(t, u(t)) \quad \text{for any} \quad i \in \mathbb{N}$$

and $t \in [s, \infty[$, that is, u is a ℓ^p-solution of (S^*_ξ). $\square$

Let us note that hypothesis $\mathbf{F^*_4}$ implies the continuity of $\tilde{F}$ (therefore in $\mathbf{F^*_3}$ the convergence is in fact uniform on every compact subset of $[s, \infty[)$. Then, it is clear that assumptions of Theorem 3.2 imply the continuity of F on $[0, \infty] \times \ell^p$ and in these hypotheses the result of Theorem 3.1 is valid.

If in (S^*_ξ) we take $\tilde{\tilde{f}}_i(t) = c_i$ for any $t \in [s, \infty[$ and $\{c_i\}_i \in \ell^p$, we obtain the autonomous case. Under this circumstance we can supplementary prove an asymptotic behaviour result.

Corollary 3.2. *Let hypotheses* $\mathbf{A_1^*}, \mathbf{A_2^*}, \mathbf{F_1^*}, \mathbf{F_2^*}$ *be valid and* $c = \alpha + \beta + \omega < 0$. *Then, there exists* $U \in \mathcal{D}$ *such that for any* $\xi \in \ell^p$, $\lim_{t \to \infty} u(t) = U$, *where* u *is the* ℓ^p- *solution of* (S_ξ^*) *with* $\left\{ \tilde{\tilde{f}}_i(t) \right\}_i = \{c_i\}_i \in \ell^p$.

Proof. According to Lemma 3.6, let $U = \{U_i\}_i \in \mathcal{D}$ be the solution of the system

$$\sum_{j=1}^{\infty} a_{ij} U_j + \tilde{f}_i(U_1, U_2, \dots) + c_i = 0, \quad i \in \mathbb{N}.$$

Hence, from Theorem 3.1 we have:

$$\|u(t) - U\| \le [\exp(\mu_A + \alpha)t] \|\xi - U\| \le (\exp ct)\|\xi - U\|.$$

From this we obtain the desired statement. $\square$

3.4. Truncation errors in linear case

Let $\xi = (\xi_i)_i \in \ell^1$ be a sequence of complex numbers and a_{ij} $(i, j = 1, 2, \dots)$ be complex valued functions defined on $[0, \infty[$ and satisfying more restrictive conditions than the above ones, namely

$\overline{\mathbf{A1}}$: a_{ij} is continuous for each $i, j = 1, 2, \dots$;

$\overline{\mathbf{A2}}$: $\sum_{i=1}^{\infty} |a_{ij}(t)| < \infty$ uniformly on every compact subset of $[0, \infty[$ for each $j = 1, 2, \dots$;

$\overline{\mathbf{A3}}$: $\alpha(t) = \sup \left\{ \sum_{i=1}^{\infty} |a_{ij}(t)| \, ; \, j = 1, 2, \dots \right\}$ is bounded on any compact subset of $[0, \infty[$.

We shall also consider functions $f_i : [0, \infty[\mapsto \mathbb{C} \, (i = 1, 2, \dots)$ with the following conditions:

$\overline{\mathbf{F1}}$: f_i is continuous for each $i = 1, 2, \dots$;

$\overline{\mathbf{F2}}$: $\sum_{i=1}^{\infty} |f_i(t)| < \infty$ uniformly on every compact subset of $[0, \infty[$.

Under these hypotheses let us consider a particular case of (S_ξ):

$$(\overline{S}_\xi) \qquad \begin{cases} \dfrac{du_i(t)}{dt} = \displaystyle\sum_{j=1}^{\infty} a_{ij}(t) u_j(t) + f_i(t), & t \ge 0, \\[2mm] u_i(0) = \xi_i, & i = 1, 2, \dots, \end{cases}$$

with the solution denoted by $u = (u_1, u_2, \dots)$ where $u(t) = (u_1(t), u_2(t), \dots) \in \ell^1$ for every $t \geq 0$. We associate the following truncated system:

$$(\overline{\mathbf{S}}_\xi^n) \qquad \begin{cases} \dfrac{du_i^n(t)}{dt} = \displaystyle\sum_{j=1}^{n} a_{ij}(t)u_j^n(t) + f_i(t), & t \geq 0, \\[4mm] u_i^n(0) = \xi_i, & i = 1, 2, \dots, n, \end{cases}$$

with the solution denoted by $(u_1^n, u_2^n, \dots, u_n^n)$. We let $u^n = (u_1^n, u_2^n, \dots, u_n^n, 0, \dots)$ and clearly $u^n(t) \in \ell^1$ for every $t \geq 0$.

When the approximation of the solution of system $(\overline{\mathbf{S}}_\xi)$ requires the calculation of the solution of the truncated system $(\overline{\mathbf{S}}_\xi^n)$ the evaluation of the truncation error $\varepsilon^n(t) = \|u(t) - u^n(t)\|_1$ is needed. An upper bound $\Delta^n(t)$ and a lower bound $\delta^n(t)$ of $\varepsilon^n(t)$ are derived below.

With the above hypotheses we can define the bounded linear operator $A(t)$: $\ell^1 \mapsto \ell^1$ with

$$A(t)u = \left\{ \sum_{j=1}^{\infty} a_{ij}(t)u_j \right\}_i ,$$

where $u = (u_j)_j \in \ell^1$. Also, let $F : [0, \infty[\mapsto \ell^1$ be a continuous function with f_i as components. It follows then immediately that $(\overline{\mathbf{S}}_\xi)$ is equivalent to the differential equation on ℓ^1:

$$(\overline{\mathbf{E}}_\xi) \qquad \begin{cases} \dfrac{du(t)}{dt} = A(t)u(t) + F(t), & t \geq 0, \\[3mm] x(0) = \xi. \end{cases}$$

Let us define a new linear bounded operator on ℓ^1, namely $A^n(t) : \ell^1 \mapsto \ell^1$ with

$$[A^n(t)u]_i = \begin{cases} \displaystyle\sum_{j=1}^{\infty} a_{ij}(t)u_j & \text{for } i = 1, 2, \dots, n, \\[4mm] \displaystyle\sum_{j=n+1}^{\infty} a_{ij}(t)u_j & \text{for } i = n+1, n+2, \dots \end{cases}$$

where $u = \{u_j\}_j \in \ell^1$ and $[A^n(t)u]_i$ is the ith component of $A^n(t)u$. We remark that if $(u_1^n, u_2^n, \dots, u_n^n)$ is the solution of $(\overline{\mathbf{S}}_\xi^n)$, then u^n satisfies the following differential equation on ℓ^1:

$$(\overline{\mathbf{E}}_\xi^n) \qquad \begin{cases} \dfrac{du^n(t)}{dt} = A^n(t)u^n(t) + F^n(t), & t \geq 0, \\[3mm] u^n(0) = \xi^n. \end{cases}$$

where $F^n(t) = (f_1(t), f_2(t), \ldots, f_n(t), 0, \ldots)$ and $\xi^n = (\xi_1, \xi_2, \ldots, \xi_n, 0, \ldots)$.

According to the continuity of the solutions of $(\overline{\mathbf{E}}_\xi)$ and $(\overline{\mathbf{E}}_\xi^n)$, we have $\sum_{i=1}^{\infty} |u_i(t) - u_i^n(t)| < \infty$ uniformly on compact subsets of $[0, \infty[$. On the other hand, let $\mathcal{M}_i$ be the countable subset of $[0, \infty[$ on which the function $t \mapsto |u_i(t) - u_i^n(t)|$ is not differentiable. On $[0, \infty[\setminus \mathcal{M}_i$ we have

$$\left| \frac{d}{dt} |u_i(t) - u_i^n(t)| \right| \leq \left| \frac{d}{dt} [u_i(t) - u_i^n(t)] \right|$$
$$\leq |[A(t)u(t)]_i| + |[A^n(t)u^n(t)]_i| + |f_i(t)| + |[F^n(t)]_i|.$$

By the continuity of $A(t)$ and $A^n(t)$ on ℓ^1 and by the assumption $\overline{\mathbf{F2}}$, the above inequality shows that $\sum_{i=1}^{\infty} \frac{d}{dt} |u_i(t) - u_i^n(t)| < \infty$ uniformly on every compact subset of $[0, \infty[\setminus \mathcal{M}$, where $\mathcal{M} = \bigcup_{i=1}^{\infty} \mathcal{M}_i$. Hence, the function $t \mapsto \|u(t) - u^n(t)\|_1 = \varepsilon^n(t)$ is differentiable on $[0, \infty[\setminus \mathcal{M}$ and (see Lemma 1.15) on this set we have

$$\frac{d\varepsilon^n(t)}{dt} = \langle u(t) - u^n(t), A(t)u(t) + F(t) - A^n(t)u^n(t) - F^n(t) \rangle_+ \tag{3.19}$$
$$= \langle u(t) - u^n(t), A(t)u(t) + F(t) - A^n(t)u^n(t) - F^n(t) \rangle_-$$

From the first part of (3.19) one obtains on $[0, \infty[\setminus \mathcal{M}$

$$\frac{d\varepsilon^n(t)}{dt} \leq \langle u(t) - u^n(t), A(t)[u(t) - u^n(t)] \rangle_+ + \tag{3.20}$$
$$+ \|A(t) - A^n(t)\|_1 \|u^n(t)\|_1 + \|F(t) - F^n(t)\|_1,$$

and from the second part

$$\frac{d\varepsilon^n(t)}{dt} \geq - \langle u(t) - u^n(t), -A(t)[u(t) - u^n(t)] \rangle_+ - \tag{3.21}$$
$$- \|A(t) - A^n(t)\|_1 \|u^n(t)\|_1 - \|F(t) - F^n(t)\|_1.$$

But, on the other hand, we can obtain, (see Lemma 3.4) for every $y = \{y_i\}_i \in \ell^1$ and $t \geq 0$,

$$\langle y, A(t)y \rangle_+ \leq \mu_A(t) \|y\|_1, \tag{3.22}$$

where

$$\mu_A(t) = \sup \left\{ \operatorname{Re} a_{jj}(t) + \sum_{\substack{i=1 \\ i \neq j}}^{\infty} |a_{ij}(t)| ; j = 1, 2, \ldots \right\}.$$

Also, we have (Coppel [1965,p.58]) for the finite system $(\overline{\mathbf{S}}_\xi^n)$

$$\|u^n(t)\|_1 = \sum_{i=1}^{n} |u_i^n(t)| \leq g_n(t), \tag{3.23}$$

where

$$g_n(t) = \left[\sum_{i=1}^{n} |\xi_i| + \int_0^t \left(\exp \int_x^0 \mu_n(s)\,ds\right) \sum_{i=1}^{n} |f_i(x)|\,dx\right] \exp \int_0^t \mu_n(s)\,ds$$

and where

$$\mu_n(s) = \sup\left\{\operatorname{Re} a_{jj}(s) + \sum_{\substack{i=1 \\ i \neq j}}^{n} |a_{ij}(s)| \,;\, j = 1, 2, \ldots, n\right\}.$$

Taking into account (3.22) and (3.23), from (3.20) we obtain on $[0, \infty[\setminus \mathcal{M}$

$$\frac{d\varepsilon^n(t)}{dt} \leq \mu_A(t)\varepsilon^n(t) + \gamma^n(t)g_n(t) + \lambda_n(t)\,, \qquad (3.24)$$

where

$$\gamma^n(t) = \|A(t) - A^n(t)\|_1 = \sup\left\{\sum_{i=n+1}^{\infty} |a_{ij}(t)| \,;\, j = 1, 2, \ldots, n\right\}$$

and

$$\lambda_n(t) = \|F(t) - F^n(t)\|_1 = \sum_{i=n+1}^{\infty} |f_i(t)|.$$

The inequality (3.24) finally gives $\varepsilon^n(t) \leq \Delta^n(t)$, where

$$\Delta^n(t) = \sum_{i=n+1}^{\infty} |\xi_i|$$
$$+ \left[\int_0^t \left(\exp \int_x^0 \mu_A(s)\,ds\right) [\gamma^n(x)g_n(x) + \lambda_n(x)]\,dx\right] \exp \int_0^t \mu_A(s)\,ds, \qquad (3.25)$$

for all $t \in [0, \infty[$.

In the same way, from (3.21) we can obtain for the lower bound $\delta^n(t)$:

$$\delta^n(t) = \sum_{i=n+1}^{\infty} |\xi_i|$$
$$- \left[\int_0^t \left(\exp \int_0^x \mu_{-A}(s)\,ds\right) [\gamma^n(x)g_n(x) + \lambda_n(x)]\,dx\right] \exp \int_t^0 \mu_{-A}(s)\,ds. \qquad (3.26)$$

Note that $\mu_{-A}(s)$ has the form of $\mu_A(s)$ with $-a_{jj}(s)$ instead of $a_{jj}(s)$.

Lemma 3.8.

(i) Under hypotheses $\overline{A1}, \overline{A2}, \overline{A3}, \overline{F1}, \overline{F2}$, for any $t \geq 0$, the sequences $\{\Delta^n(t)\}_n$ and $\{\delta^n(t)\}_n$ are bounded.

(ii) If $\overline{A1}, \overline{A2}, \overline{A3}, \overline{F1}, \overline{F2}$ are assumed together with the hypothesis

$\overline{A4}$: $\{\gamma^n(t)\}_n \to 0$ uniformly on every compact subset of $[0, \infty[$,

then $\{\Delta^n(t)\}_n$ and $\{\delta^n(t)\}_n$ converge to 0 for any $t \geq 0$.

Proof. The lemma follows directly from the inspection of expressions (3.25) and (3.26) and taking into account the fact that for any n and t, $\mu_n(t) \leq \mu_A(t) \leq \alpha(t), \gamma_n(t) \leq \alpha(t)$ and $\lambda_n(t) \leq \|F(t)\|_1$. $\square$

Lemma 3.9. *Assume the initial hypotheses* $\overline{A1}, \overline{A2}, \overline{A3}, \overline{F1}, \overline{F2}$ *together with the hypotheses:*

$\overline{A5}$:   For any positive integer n, there is $M_n \geq 0$ such that $\gamma^n(t) \leq M_n$ for all $t \geq 0$.

$\overline{A6}$:   There is a strictly positive number ν such that

$$\operatorname{Re} a_{jj}(t) + \sum_{\substack{i=1 \\ i \neq j}}^{\infty} |a_{ij}(t)| \leq -\nu < 0 \text{ for any } j = 1, 2, \ldots \text{ and any } t \in [0, \infty[.$$

$\overline{F3}$:   The function $t \mapsto F(t)$ is bounded on $[0, \infty[: \|F(t)\|_1 \leq N$.

Then, the function $t \mapsto \Delta^n(t)$ is bounded for any $n = 1, 2, \ldots$

Proof. Hypothesis $\overline{A6}$ implies

$$0 < \frac{1}{-\mu_n(t)} \leq \frac{1}{-\mu_A(t)} \leq \nu^{-1} .$$

Denoting

$$\tilde{L}_n(t) = \sup\left\{ \sum_{i=1}^{n} |f_i(x)| \; : \; 0 \leq x \leq t \right\},$$

it follows that

$$g_n(x) \leq \left[\sum_{i=1}^{n} |\xi_i| - \nu^{-1} \tilde{L}_n(t) \right] \exp \int_0^x \mu_n(s) \, ds + \nu^{-1} \tilde{L}_n(t),$$

for any $x \in [0, t]$. Thus we derive the inequality

$$\int_0^t \gamma^n(x) g_n(x) \left(\exp \int_x^0 \mu_A(s) \, ds \right) dx \leq$$

$$\leq M_n \sum_{i=1}^n |\xi_i| t + \frac{M_n \tilde{L}_n(t)}{\nu^2} \left(\exp \int_t^0 \mu_A(s) \, ds - 1 \right). \quad (3.27)$$

We also have that

$$\int_0^t \lambda_n(x) \left(\exp \int_x^0 \mu_A(s) \, ds \right) dx \leq \nu^{-1} \tilde{\tilde{L}}_n(t) \left(\exp \int_t^0 \mu_A^1(s) \, ds - 1 \right), \quad (3.28)$$

where

$$\tilde{\tilde{L}}_n(t) = \sup \left\{ \sum_{i=n+1}^\infty |f_i(x)| \; : \; x \in [0, t] \right\}.$$

By using (3.27) and (3.28) it follows that

$$\Delta^n(t) \leq \left(\sum_{i=n+1}^\infty |\xi_i| + M_n t \sum_{i=1}^n |\xi_i| \right) \exp(-\nu t) + M_n \nu^{-2} \tilde{L}_n(t) + \nu^{-1} \tilde{\tilde{L}}_n(t). \quad (3.29)$$

The assertions of the Lemma 3.9 follows from (3.29), as $\tilde{L}_n(t), \tilde{\tilde{L}}_n(t) \leq N$. □

Expression (3.29) constitutes an upper bound for the truncation error (valid under the assumptions given in the statement of the lemma), which are easier to apply than (3.25).

3.5. Applications to infinite circuits

Let us now verify that the theory developed above really works for the network in Figure 3.1 described by (3.5). In this case we have constant coefficients and $p = 1$. By (3.1), (3.2) and the continuity of functions f_j we see that the

hypotheses $\mathbf{A_1^*, A_2^*, F_1^*, F_3^*}$ hold with $\beta = \sup \left\{ \sum_{\substack{k=1 \\ k \neq j}}^\infty |G_{kj}|/c_j \; ; \; j \geq 2P + 1 \right\}$ and

$\omega = \sup \{ -G_{jj}/c_j \; ; \; j \in \mathbb{N} \}$.

Let us denote

$$\alpha^j = \begin{cases} \alpha_f^{\frac{j+1}{2}} & j \in \{1, 3, \ldots, 2P - 1\} \\ \alpha_r^{\frac{j}{2}} & j \in \{2, 4, \ldots, 2P\} \end{cases}$$

and

$$\alpha = \max_{1 \leq j \leq 2P} \left[\frac{G_{jj}}{c_j} + \max \left(\frac{-1 + \alpha^j}{\tau_j} \; ; \; \frac{-G_{jj} + \sum\limits_{\substack{k=1 \\ k \neq j}}^{\infty} |G_{kj}|}{c_j} \right) \right] \geq 0.$$

Let us consider $z^1 = \{z^1_j\}_j$ and $z^2 = \{z^2_j\}_j$ both in ℓ^1. Then, the mean value theorem yields

$$\tilde{f}_k(z^1_1, z^1_2, \ldots) - \tilde{f}_k(z^2_1, z^2_2, \ldots) = \frac{\delta_k G_{kk}}{c_k}(z^1_k - z^2_k) - \sum_{j=1}^{2P} \frac{t_{kj} f'_j(r_j) + G_{kj}}{\tau_j f'_j(r_j) + c_j}(z^1_j - z^2_j) \, ,$$

where $r_j = \gamma_j^{-1}(z_j)$ and $z_j \in]z^1_j, z^2_j[$. If we denote $Y_0 = \{k \in \mathbf{N} \; ; \; z^1_k - z^2_k = 0\}$, then in view of Lemma 1.7 we have

$$\langle z^1 - z^2, \tilde{F}(z^1) - \tilde{F}(z^2) \rangle_+$$

$$= \sum_{k \in Y_0} | \sum_{j=1}^{2P} \frac{-t_{kj} f'_j(r_j) - G_{kj}}{\tau_j f'_j(r_j) + c_j}(z^1_j - z^2_j) + \frac{\delta_k G_{kk}}{c_k}(z^1_k - z^2_k) | +$$

$$+ \sum_{k \notin Y_0} \frac{|z^1_k - z^2_k|}{z^1_k - z^2_k} \left[\sum_{j=1}^{2P} \frac{-t_{kj} f'_j(r_j) - G_{kj}}{\tau_j f'_j(r_j) + c_j}(z^1_j - z^2_j) + \frac{\delta_k G_{kk}}{c_k}(z^1_k - z^2_k) \right]$$

$$\leq \sum_{k=1}^{2P} |z^1_k - z^2_k| \left[\frac{-t_{kk} f'_k(r_k) - G_{kk}}{\tau_k f'_k(r_k) + c_k} + \frac{G_{kk}}{c_k} \right] + \sum_{k=1}^{\infty} \sum_{\substack{j=1 \\ j \neq k}}^{2P} \frac{|t_{kj}| f'_j(r_j) + |G_{kj}|}{\tau_j f'_j(r_j) + c_j} |z^1_j - z^2_j|$$

$$= \sum_{j=1}^{2P} |z^1_j - z^2_j| \left[\frac{G_{jj}}{c_j} + \frac{\left(-t_{jj} + \sum\limits_{\substack{k=1 \\ k \neq j}}^{\infty} |t_{kj}| \right) f'_j(r_j) - \left(G_{jj} + \sum\limits_{\substack{k=1 \\ k=j}}^{\infty} |G_{kj}| \right)}{\tau_j f'_j(r_j) + c_j} \right]$$

$$\leq \alpha \sum_{j=1}^{2P} |z^1_j - z^2_j|,$$

because

$$-t_{jj} + \sum_{\substack{k=1 \\ k \neq j}}^{\infty} |t_{kj}| = -1 + \alpha^j.$$

This is the total dissipativity of $\tilde{F} - \alpha \mathcal{J}$ on ℓ^1, i.e. hypothesis $\mathbf{F}^*_2$ is valid. Finally, we add a new assumption for our circuit, namely there is a continuous function d :

$[0, \infty[^2 \mapsto [0, \infty[$ such that for any $t, s \in [0, \infty[$, $\sum\limits_{j=-n+1}^{0} |e_j(t) - e_j(s)| \leq |t - s| d(t, s)$.

With this very reasonable hypothesis about the circuit sources, it is clear that $\mathbf{F_4^*}$ holds. Thus, from Theorems 3.2 and 3.1 we obtain the existence and uniqueness of $q = \{q_j\}_j \in C^1_{[0,\infty[}$, ℓ^1-solution of (3.5). By considering

$$\mu_A = \sup \left\{ -\frac{G_{jj}}{c_j} + \sum_{\substack{k=1 \\ k \neq j}}^{\infty} |G_{kj}| / c_j \; ; \; j \in \mathbb{N} \right\},$$

Corollary 3.1 gives stability conditions related to the diagonal dominance of the matrix G, as in the finite dimensional case treated in the previous chapter.

Let us consider now a concrete example of an infinite linear circuit. It originates from an infinite length distributed structure described by Telegraph Equations (Ghausi and Kelly [1968]):

$$\begin{cases} \dfrac{\partial u(t, x)}{\partial x} = -r(x) i(t, x) \\ \dfrac{\partial i(t, x)}{\partial x} = -c(x) \dfrac{\partial u(t, x)}{\partial t} - g(x) u(t, x) \\ t \geq 0, \quad x \geq 0 \, . \end{cases}$$

Here, $u(t, x)$ and $i(t, x)$ are the voltage and the current respectively, at the moment t and at the point x along the line. The distributed parameters (per unit length) will be the conductance $g(x) = g$ and the capacitance $c(x) = c$, both constant and positive, and the resistance $r(x) > 0$ a strictly increasing function.

By discretization of the above equations with respect to the x-variable only, and with a constant step $h > 0$ we obtain:

$$\begin{cases} \tilde{u}_{j+1} - \tilde{u}_j = -hr(hj)\tilde{i}_j, \quad j = 0, 1, 2, \ldots \\ \tilde{i}_j - \tilde{i}_{j-1} = -hc\dfrac{d\tilde{u}_j}{dt} - hg\tilde{u}_j, \quad j = 1, 2, \ldots \end{cases} \tag{3.30}$$

Above, we have denoted by $\tilde{u}_j$ and $\tilde{i}_j$ the approximation of $u(t, x)$ and $i(t, x)$ respectively, at the point $x = x_j = hj$, $j = 0, 1, 2, \ldots$

If the line is excited by the constant source $e_0 = \tilde{u}_0$ at the left hand end, we obtain the infinite ladder network illustrated in Figure 3.2, where $r_j = r(hj)$, $j = 0, 1, 2, \ldots$

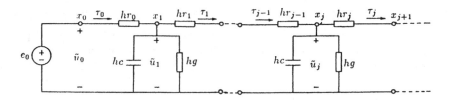

Figure 3.2 Infinite ladder network

This network is of the form of that in Figure 3.1, with $n = 1$ and $P = 0$. The system governing this circuit is derived from (3.30), being of the form $\overline{S}_\xi$ with constant coefficients, namely

$$
\begin{cases}
\dfrac{d\tilde{u}_j}{dt} = \dfrac{\tilde{u}_{j-1}}{ch^2 r_{j-1}} - \tilde{u}_j \left(\dfrac{1}{ch^2 r_j} + \dfrac{1}{ch^2 r_{j-1}} + \dfrac{g}{c} \right) + \dfrac{\tilde{u}_{j+1}}{ch^2 r_j} \\[2ex]
\tilde{u}_0 = e_0 \\[1ex]
\tilde{u}_j(0) = 0, \quad j = 1, 2, \ldots
\end{cases}
$$

All conditions for the existence of a unique uniform asymptotically stable solution $\tilde{u}_j = q_j/(ch) \in \ell^1$ are fulfilled. In fact $\overline{\textbf{A1}} - \overline{\textbf{A6}}$ and $\overline{\textbf{F1}} - \overline{\textbf{F3}}$ are valid with $\mu_n = \mu_A = -\nu = -g/c$, $\gamma^n = 1/(h^2 r(hn)c)$, $\lambda_n = 0$ for $n > 1$. Clearly, it is interesting to evaluate the error when we keep n-cells of the ladder network. If zero initial conditions are considered, then the error is bounded from above by (see (3.25)):

$$
\Delta^n(t) = \frac{e_0}{h^4 r(hn) r(0) gc} \left(\frac{c}{g} - \frac{c}{g} e^{-(g/c)t} - t e^{-(g/c)t} \right)
$$

and bounded from below by (see (3.26)):

$$
\delta^n(t) =
$$

$$
\frac{e_0}{h^4 r(hn) r(0) gc} \left\{ \mu_{-A}^{-1} + \left[\left(\mu_{-A} - \frac{g}{c} \right)^{-1} - \mu_{-A}^{-1} \right] e^{-\mu_{-A} t} - \left(\mu_{-A} - \frac{g}{c} \right)^{-1} e^{-(g/c)t} \right\}
$$

where

$$
\mu_{-A} = \max \left(\frac{2}{r_1 ch^2} + \frac{1}{r_0 ch^2} + \frac{g}{c} \ ; \ \frac{2}{r_1 ch^2} + \frac{2}{r_2 ch^2} + \frac{g}{c} \right).
$$

Because $\lim_{n \to \infty} \Delta^n(t) = \lim \delta^n(t) = 0$, the solution of the truncated circuit converges to the one of the infinite network. The above bounds can be useful to choose adequate values of the number of cells and of the step size with the view of obtaining a desired error. Let us finally remark the boundedness of functions $t \mapsto \Delta^n(t)$ and $t \mapsto \delta^n(t)$.

Chapter IV

Mixed-type circuits with distributed and lumped parameters as correct models for integrated structures

4.0. Why mixed-type circuits?

The technology of integrated circuits imposes upon their designers the need to deal with structures with distributed parameters. Figure 4.1 shows a schematic diagram of part of a digital integrated chip, consisting of an n MOS transistor with gate (G), drain (D) and source (S) as terminals, and its thin-film connection with the rest of the chip. This on-chip connection can be made by metals (Al, W), polycristaline silicon (polysilicon) or metal silicides (WSi_2). Alternative materials to oxide-passivated silicon substrates are saphire and gallium arsenide (Saraswat and Mohammadi [1982], Yuan et al. [1982], Passlack et al. [1990]).

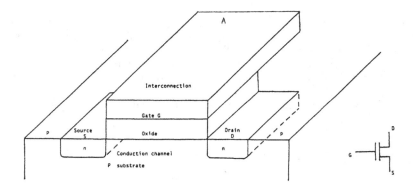

Figure 4.1 Cross section of MOS transistor with interconnection

The performance criteria for such a digital circuit are: a high operating speed, a high level of integration, a small chip area, the nonexistence of false switching and low power consumption. The operating speed is given by the clock frequency, in close relation with the rise time. Referring to Figure 4.1, "the rise time", for this circuit, is the time required for the output voltage (the drain potential, for example) to rise (or fall) from its initial value to 90 percent of its final value under a step-voltage located at point A – the output of another stage (although different by definition, our "delay time" used in Chapter V will express also "the inertia" of the circuit). Primary interest at the design stage is to be able to predict this performance which is clearly dependent on the delays caused both by devices and by interconnections. But, with the advances in technology, the cross section of connecting wires decreases while its length increases as a result of the increase in the number of devices on the same chip (integration scale). That is why the delay time associated with interconnections becomes an appreciable part of the total delay time. In certain cases when the wiring lengths are as short as 1 mm with 4 μm minimum feature size, or for recent advanced $GaAs$ MESFET and $GaAs$ HEMT technologies, the interconnection delay dominates the global delay (Saraswat and Mohammadi [1982], Bakoglu and Meindl [1985]).

Our goal is to study the delay caused by connecting wires. Then, it is reasonable to consider the simplest models for MOS transistors: a resistor between drain and source having a very small resistance in the ON state and a very high one in the OFF state. Capacitances associated with the pull-up source diffusion, contact cuts and the gates being driven can be included by connecting respective nodes to the ground. Our model is not restricted to MOS circuits at all. The bipolar circuits where grounded resistors can appear (O'Brien and Wyatt [1986]) are also approachable.

Naturally, being interested in wiring delay, we try to model the interconnections as exact as possible. The required frequencies make valid the quasi-transverse electromagnetic wave approximation (Wohlers [1969]). At the same time, with subnanosecond rise times, the electrical length of interconnection can become a significant fraction of the wave length. That is why the transmission line property of these interconnections can no longer be neglected if we desire an accurate model and therefore we shall use the well-known Telegraph Equations (Ghaussi and Kelly [1968]).

$$(TE) \qquad \begin{cases} \dfrac{\partial v}{\partial x} = -ri \\ \dfrac{\partial i}{\partial x} = -c\dfrac{\partial v}{\partial t} - gv \ . \end{cases}$$

Here $v(t, x)$ and $i(t, x)$ are respectively the voltage and the current at the moment t at a point x on the distributed structure ("rcg-line"). The distributed parameters (all per unit length) are: resistance of conductive path $r > 0$, capacitance $c > 0$

and conductance $g \geq 0$ of the dielectric substrate (SiO_2 in Figure 4.1). We have neglected the distributed inductance of the conductive layer, reducing the frequency range of our model validity some tens of megahertz. Thus our results cannot be applied to microwave circuits but are valid for on-chip and inter-chip interconnections for most digital systems.

The mathematical model resulting from the above considerations is a system of partial differential equations coupled by the boundary conditions which imply ordinary differential equations. The following section explains, by means of examples, the relevance of the well-possedness problem for our model. This model is precisely formulated in Section 4.2. The rest of the chapter gives conditions under which there exists a solution (in a well-precised sense) of the direct current (steady-state) or dynamic regime of our circuit model. We work in a space of the form $L_2 \times \mathbf{R}$ and the central fact is again the dissipativity property. Conditions assuming the uniqueness of the solution as well as the continuous dependence on sources and initial data are also given. It has been found that the (semi)positivity of the matrix describing the lumped resistive part of the network is the basic requirement for the model correctness. Also, the strengthening of the hypotheses gives an asymptotic stability property.

4.1. Examples

Two examples below show that some mixed models can be wrongly formulated.

Let us consider the circuit presented in Figure 4.2 which contains an inverter formed by T_1, T_2. It controls the gates T_4, T_5 through the lines L_1 and L_2 and the pass transistor T_3.

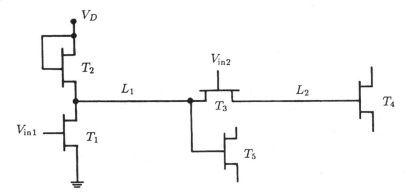

Figure 4.2 A digital circuit example

According to the modelling procedure described above, we substitute low and high drain to source resistance for "active" T_1 and T_3 transistors in the "on" and "off" states respectively. The "on" and "off" state are controlled by voltages $V_{\text{in}1}$ and $V_{\text{in}2}$ respectively. By modelling transistors T_4 and T_5 by RC lumped parameter equivalent circuits, one obtains the circuit presented in Figure 4.3.

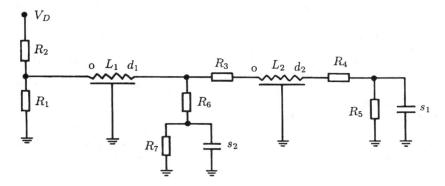

Figure 4.3 A possible mixed model for Figure 4.2

Suppose that transistor T_3 is continuously in the on-state when studying the dynamic process associated with commuting the transistor T_1 from the on state (i.e. resistance R_1' for $t \le 0$) to the off state (i.e. $R_1 \gg R_1'$ for $t > 0$). Let us denote by $v_{10}(x)$ the voltage across the line L_1 during the initial steady state; it is then trivial to derive (by using system (TE)) the following relation at the left hand end $(x = 0)$ of this line:

$$V_D = R_2 \left[-\frac{1}{r_1} \frac{\partial v_{10}}{\partial x}(0) + \frac{v_{10}}{R_1'}(0) \right] + v_{10}(0). \tag{4.1}$$

If $v_1(t, x)$ denotes the voltage across L_1 during the studied dynamic regime we have

$$V_D = R_2 \left[-\frac{1}{r_1} \frac{\partial v_1}{\partial x}(t, 0) + \frac{v_1(t, 0)}{R_1} \right] + v_1(t, 0). \tag{4.2}$$

Equations (4.1) and (4.2) show that the mixed model of the circuit does not possess a solution in the "classical sense" (i.e. a continuously differentiable function in the x and t variables) under the initial condition $v_{10}(x)$. Indeed, if such a solution is supposed to exit, then we have to obtain

$$\lim_{t \to 0} v_1(t, 0) = v_{10}(0)$$

and

$$\lim_{t \to 0} \frac{\partial v_1}{\partial x}(t,0) = \frac{\partial v_{10}}{\partial x}(0),$$

contradicting (4.1) and (4.2), as $R'_1 \neq R_1$. We have encountered here a case of what mathematicians call inconsistency, that is, initial conditions which do not satisfy the boundary conditions. It is then necessary to search for generalized solutions (or solutions in "the sense of distributions") instead of "classical sense" ones. Moreover, we have to accept generalized solutions in the case of discontinuous sources. It was noted long before (Fattorini [1983]) that the classical solutions "are in no way required by nature and the generalized solutions are perfectly acceptable when modelling physical phenomena."

Let us observe that in the above example "the inconsistency" was required by the simplest model of the transistor T_1 considered as an ideal switch with abrupt passing from R'_1 to R_1 drain to source resistance. If we consider a more elaborate model for T_1, then in our mathematical model the initial conditions would satisfy the boundary ones but, of course, the equations would be much more complicated.

Another example is the simple circuit presented in Figure 4.4.

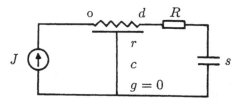

Figure 4.4 A network without steady state

The direct current state of this circuit is described by equation

$$\frac{d^2 v}{dx^2} = 0$$

under boundary conditions

$$-\frac{1}{r}\frac{dv}{dx}(0) = J \quad \text{and} \quad -\frac{1}{r}\frac{dv}{dx}(d) = 0.$$

It is easy to see that such a problem has no solution in a classical sense if $J \neq 0$. It can be shown, Mikhailov [1980,p.190], that the problem does not even have a generalized solution, so that the proposed model is improper.

4.2. Statement of the problem

We study the general network presented in Figure 4.5.

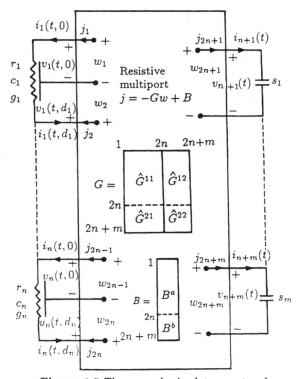

Figure 4.5 The general mixed–type network

The linear lumped resistive part of the network is concentrated in a so called $(G, B, 2n + m)$–multiport. This means that, if j and w are the vectors of currents and voltages at the $2n + m$–ports, then we have

$$j = -Gw + B(t) \qquad (4.3)$$

where G is a $2n + m$–order square matrix of conductances, while $B(t)$ is a $2n + m$ vector with elements being linear combinations of independent sources. (An example to establish G and $B(t)$ in a concrete case can be found in Section 6.7.)

The first $2n$ terminals of the multiport are connected to n elements with parameters distributed over their entire length: resistance $r_i > 0$, capacitance $c_i > 0$ and conductance $g_i \geq 0$, $i = \overline{1,n}$. Linear capacitors (with capacitance $s_i > 0$, $i = \overline{1,m}$) are connected to the last m ports.

Of course there exist resistive networks with $2n + m$ pairs of terminals which are not of type $(G, B, 2n + m)$. For instance, the following conditions imposed on our mixed-type circuit assure the existence of the $(G, B, 2n + m)$-type description of its resistive part:

- one of the multiport terminals is common for all external elements (this is the common "ground" of the rcg-lines and of the capacitors),
- all sources are independent,
- none of the 0 or d_i terminals is connected directly (i.e. through a zero–resistance branch) to the ground,
- there is no direct connection between 0 and/or d_i-terminals. Also, two or more terminals with capacitors (which are not "the ground") are not connected directly together.

Indeed, let us consider the resistive network with all sources removed. We may simplify this network by using the star-mesh transformation to remove all internal nodes. Clearly we obtain a circuit described by $j = -Gw$. Moreover, the matrix G is symmetric and "weakly diagonally row sum dominant" (WDRD). This means that, for all $i = \overline{1, 2n + m}$ we have

$$G_{ii} \geq S_i = \sum_{\substack{j=1 \\ j \neq i}}^{2n+m} |G_{ij}| \, , \tag{4.4}$$

where G_{ij} are elements of the matrix G. Clearly a WDRD matrix is semi-positive definite $G \geq 0$. Also, for many networks in the subclass considered above, G is "diagonally row sum dominant" (DRD):

$$G_{ii} > S_i = \sum_{\substack{j=1 \\ j \neq i}}^{2n+m} |G_{ij}| \tag{4.5}$$

for all $i = \overline{1, 2n + m}$. Of course, the DRD condition implies the positivity of G.

For an arbitrary but fixed $T > 0$ let us denote by $v_k :]0, T[\times]0, d_k[\rightarrow \mathbf{R}$, the voltage along the kth distributed element and by $i_{n+k} :]0, T[\rightarrow \mathbf{R}$, the kth capacitor current. We observe in Figure 4.5 that

$$v_k(t, 0) = w_{2k-1}, \ v_k(t, d_k) = w_{2k},$$
$$i_k(t, 0) = j_{2k-1}, \ i_k(t, d_k) = -j_{2k}$$

for $k = \overline{1,n}$ and that

$$v_{n+k}(t) = w_{2n+k},$$
$$i_{n+k}(t) = j_{2n+k}$$

for $k = \overline{1,m}$.
From (TE) we formally derive the system:

$$(E) \qquad \frac{\partial v_k}{\partial t} = \frac{1}{r_k c_k} \frac{\partial^2 v_k}{\partial x^2} - \frac{g_k}{c_k} v_k, \quad t \in \,]0,T[,\ x \in \,]0,d_k[,\ k = \overline{1,n}.$$

Also, from (4.3) and (TE) we obtain the following system of boundary conditions:

$$(BC) \qquad
\begin{bmatrix}
-\dfrac{1}{r_1}\dfrac{\partial v_1}{\partial x}(t,0) \\[2mm]
+\dfrac{1}{r_1}\dfrac{\partial v_1}{\partial x}(t,d_1) \\[2mm]
\vdots \\[2mm]
-\dfrac{1}{r_n}\dfrac{\partial v_n}{\partial x}(t,0) \\[2mm]
+\dfrac{1}{r_n}\dfrac{\partial v_n}{\partial x}(t,d_n) \\[2mm]
s_1 \dfrac{dv_{n+1}}{dt}(t) \\[2mm]
\vdots \\[2mm]
s_m \dfrac{dv_{n+m}}{dt}(t)
\end{bmatrix}
= -G
\begin{bmatrix}
v_1(t,0) \\[2mm]
v_1(t,d_1) \\[2mm]
\vdots \\[2mm]
v_n(t,0) \\[2mm]
v_n(t,d_n) \\[2mm]
v_{n+1}(t) \\[2mm]
\vdots \\[2mm]
v_{n+m}(t)
\end{bmatrix}
+ B(t), \quad t \in \,]0,T[\,,$$

where the capacitor equations $i_{n+k}(t) = s_k \frac{dv_{n+k}(t)}{dt}$ were also used. Finally, adding
the vector of initial conditions $v_0 = (v_{1,0}, \ldots, v_{n,0}, \ldots, v_{n+m,0})$ we obtain

$$(IC) \qquad
\begin{cases}
v_k(0,x) = v_{k,0}(x) & ;\ x \in \,]0,d_k[,\ k = \overline{1,n} \\[2mm]
v_{n+k}(0) = v_{n+k,0} & ;\ k = \overline{1,m}\,.
\end{cases}$$

Thus, our dynamic problem will be

$$(P(B,v_0)) = (E) + (BC) + (IC).$$

At the same time we shall deal with the time independent (or steady-state, or direct
current) problem corresponding to $(P(B,v_0))$ and denoted by $(SP(B))$. It can be
obtained from (E) and (BC) by taking $B(t) = B = $ constant, and cancelling the
time derivates. So,

$$(SP(B)) = (SE) + (SBC),$$

where

(SE) $\qquad \dfrac{d^2 v_k}{dx^2} - g_k r_k v_k = 0 \quad ; \ x \in \ ]0, d_k[, \ k = \overline{1, n}$

and

(SBC)

$$
\begin{bmatrix}
-\dfrac{1}{r_1}\dfrac{dv_1}{dx}(0) \\[2mm]
+\dfrac{1}{r_1}\dfrac{dv_1}{dx}(d_1) \\[2mm]
\vdots \\[2mm]
-\dfrac{1}{r_n}\dfrac{dv_n}{dx}(0) \\[2mm]
+\dfrac{1}{r_n}\dfrac{dv_n}{dx}(d_n) \\[1mm]
0 \\
\vdots \\
0
\end{bmatrix}
= -G
\begin{bmatrix}
v_1(0) \\
v_1(d_1) \\
\vdots \\
v_n(0) \\
v_n(d_n) \\
v_{n+1} \\
\vdots \\
v_{n+m}
\end{bmatrix}
+ B \ .
$$

Let us now present some notations that are used throughout this chapter. Let $\mathbf{K}$ be the set of the real ($\mathbf{K} = \mathbf{R}$) or complex ($\mathbf{K} = \mathbf{C}$) numbers. $\mathbf{K}_s^m$ is the space $\mathbf{K}^m$ with the euclidean norm weighted by positive constants $s_1, \ldots, s_m$. If a square matrix is semipositive (positive) definite we shall put $G \geq 0 \ \ (G > 0)$; G^{tr} is the transpose of G.

For $T \in \ ]0, \infty[$, m a positive integer and X a normed space, $C^m(0, T; X)$ will be the space of functions, defined on $]0, T[$, with values in X and with continuous derivatives (up to and including order m). Let $C_0^\infty(0, d_i; \mathbf{K})$ be the functions from $C^\infty(0, d_i; \mathbf{K})$ with compact support. For $\nu \in]0, 1]$, $C^\nu(0, T; X)$ will denote the space of functions for which there exists $M > 0$ such that $\|f(t_1) - f(t_2)\| \leq M|t_1 - t_2|^\nu$ for all $t_1, \ t_2 \in \ ]0, T[$ (Hölder continuity);

$$
C^{1+\nu}(0, T; X) = \left\{ f \in C^1(0, T; X) \ : \ \frac{df}{dt} \in C^\nu(0, T; X) \right\}.
$$

By $L_1(0, T; X)$ we denote the space of integrable functions and by $L_{2,i}(0, d_i; \mathbf{K}) \equiv L_{2,i}$ the measurable functions with $\|f\|_{L_{2,i}}^2 = \int_0^{d_i} c_i |f(x)|^2 \, dx < \infty$, and with the scalar product $\langle f, g \rangle_{L_{2,i}} = \int_0^{d_i} c_i f(x) \overline{g(x)} \, dx$. Let also $L_2^n(\mathbf{K}) = \prod_{i=1}^n L_{2,i} \equiv L_2^n$ with $\|f\|_{L_2^n}^2 = \sum_{i=1}^n \|f_i\|_{L_{2,i}}^2$ where f_i are the components of f. $H_{m,i}$ is the Sobolev space of functions from $L_{2,i}$ with generalized (or distributional) derivatives (up to and including order m) also in $L_{2,i}$; $H_m^n(\mathbf{K}) = \prod_{i=1}^n H_{m,i} \equiv H_m^n$.

4.3. Existence and uniqueness for dynamic system

In the sequel, it is convenient that the first $2n$ equations from (BC) are homogeneous. This is why we shall take

$$v_k = u_k + u_k^*, \quad k = \overline{1, m+n} , \tag{4.6}$$

where

$$\begin{cases} u_k^*(t, x) = \alpha_k(t)x^3 + \beta_k(t)x^2 + \delta_k(t)x, & k = \overline{1, n} \\ u_k^*(t, x) = 0, & k = \overline{n+1, n+m}. \end{cases} \tag{4.7}$$

If we denote by $b_k(t)$, $k = \overline{1, 2n+m}$ the components of $B(t)$, the coefficients in (4.7) are fixed by the conditions

$$\begin{cases} u_k^*(t, 0) = u_k^*(t, d_k) = 0 \\ -\dfrac{1}{r_k}\dfrac{\partial u_k^*}{\partial x}(t, 0) = b_{2k-1}(t) \\ \dfrac{1}{r_k}\dfrac{\partial u_k^*}{\partial x}(t, d_k) = b_{2k}(t) \end{cases} \tag{4.8}$$

such that it follows

$$\begin{cases} \alpha_k(t) = r_k d_k^{-2}[b_{2k}(t) - b_{2k-1}(t)] \\ \beta_k(t) = r_k d_k^{-1}[2b_{2k-1}(t) - b_{2k}(t)] \\ \delta_k(t) = -r_k b_{2k-1}(t) \end{cases} \tag{4.9}$$

for all $k = \overline{1, n}$. To simplify the writing below, the following notations will be used for $k = \overline{1, n}$

$$\tilde{b}_k(t, x) = -\left[\frac{d\alpha_k}{dt} + \frac{g_k}{c_k}\alpha_k(t)\right]x^3 - \left[\frac{d\beta_k}{dt} + \frac{g_k}{c_k}\beta_k(t)\right]x^2$$
$$-\left[\frac{d\delta_k}{dt} + \frac{g_k}{c_k}\delta_k(t) - \frac{6\alpha_k(t)}{r_k c_k}\right]x + \frac{2\beta_k}{r_k c_k} \tag{4.10}$$

and

$$\tilde{B}^a = (\tilde{b}_1, \ldots, \tilde{b}_n)^{tr}, \quad \tilde{B}^b = \left(\operatorname{diag}\frac{1}{s_k}\right)(b_{2n+1}, \ldots, b_{2n+m})^{tr}$$
$$u^a = (u_1, \ldots, u_n)^{tr}, \quad u^b = (u_{n+1}, \ldots, u_{n+m})^{tr}$$

and similarly for u^{*a} and u^{*b}, v^a and v^b.

Also, let us denote

$$(\gamma_0 u^a)(t) = (u_1(t,0), u_1(t,d_1), \ldots, u_n(t,0), u_n(t,d_n))^{tr}$$

$$(\gamma_1 u^a)(t) = \left(-\frac{1}{r_1}\frac{\partial u_1}{\partial x}(t,0), \frac{1}{r_1}\frac{\partial u_1}{\partial x}(t,d_1), \ldots, \right.$$

$$\left. , -\frac{1}{r_n}\frac{\partial u_n}{\partial x}(t,0), \frac{1}{r_n}\frac{\partial u_n}{\partial x}(t,d_n) \right)^{tr}$$

$$N = \operatorname{diag}\frac{1}{r_k c_k}, \qquad P = \operatorname{diag}\frac{g_k}{c_k},$$

$$\tilde{G}_{21} = \left(\operatorname{diag}\frac{1}{s_k} \right)\hat{G}_{21}, \quad \tilde{G}_{22} = \left(\operatorname{diag}\frac{1}{s_k} \right)\hat{G}_{22}$$

where $\hat{G}_{21}, \hat{G}_{22}$ together with $\hat{G}_{11}$ and $\hat{G}_{12}$ are block matrices composing G (see Figure 4.5).

With these notations, and after a change of variables for (4.6), the problem $(P(B, v_0)) = (E) + (BC) + (IC)$ becomes

(E^1)
$$\frac{\partial u^a}{\partial t} = N\frac{\partial^2 u^a}{\partial x^2} - Pu^a(t,x) + \tilde{B}^a(t,x)$$

(BC^1)
$$\begin{cases} (\gamma_1 u^a)(t) = -\hat{G}_{11}(\gamma_0 u^a)(t) - \hat{G}_{12} u^b(t) \\ \dfrac{du^b}{dt} = -\tilde{G}_{21}(\gamma_0 u^a)(t) - \tilde{G}_{22} u^b(t) + \tilde{B}^b(t) \end{cases}$$

(IC^1)
$$\begin{cases} u^a(0,x) = v^a(0,x) - u^{*a}(0,x) \\ u^b(0) = v^b(0) - u^{*b}(0) . \end{cases}$$

Of course, the above derivatives are in componentwise meaning.

Let $X_{\mathbf{K}} = L_2^n(\mathbf{K}) \times \mathbf{K}_s^m$ the Hilbert space endowed with the inner product

$$\langle f, g \rangle_{X_{\mathbf{K}}} = \langle f^a, g^a \rangle_{L_2^n(\mathbf{K})} + \langle f^b, g^b \rangle_{\mathbf{K}_s^m} =$$

$$= \sum_{1=1}^n \int_0^{d_i} c_i f_i(x)\overline{g_i(x)}\, dx + \sum_{i=n+1}^{n+m} s_{i-n} f_i \overline{g_i} .$$

We consider the operator A with the domain

$$\mathcal{D}(A) = \left\{ u \in X_{\mathbf{K}};\ u = \begin{bmatrix} u^a \\ u^b \end{bmatrix} \in H_2^n(\mathbf{K}) \times \mathbf{K}_s^m,\ \gamma_1 u^a = -\hat{G}_{11}\gamma_0 u^a - \hat{G}_{12} u^b \right\}$$

and with the definition

$$Au = \begin{bmatrix} N\dfrac{d^2u^a}{dx^2} - Pu^a \\ -\tilde{G}_{21}\gamma_0 u^a - \tilde{G}_{22}u^b \end{bmatrix} .$$

Here $\gamma_0 u^a$ and $\gamma_1 u^a$ have the same definitions as before except the time variable is absent. Then, the problem $(E^1) + (BC^1) + (IC^1)$ suggests that we formulate the following Cauchy problem on the space $X_{\mathbf{R}}$:

$(CP(\tilde{B}, u_0))$
$$\begin{cases} \dfrac{du}{dt} = Au + \tilde{B}(t) \\ u(0) = u_0 \equiv v(0) - u^*(0) \in X_{\mathbf{R}} , \end{cases}$$

where

$$\tilde{B}(t) = \begin{bmatrix} \tilde{B}^a(t) \\ \tilde{B}^b(t) \end{bmatrix} .$$

In view of studying the problem $(CP(\tilde{B}, u_0))$, the following two lemmas give properties of operator A.

Lemma 4.1. If $G \geq 0$ $(G > 0)$ then A is dissipative (strongly dissipative) in $X_{\mathbf{K}}$.

Proof. We have to prove that if $G \geq 0$, then

$$\mathrm{Re}\langle Au, u\rangle_{X_{\mathbf{K}}} \leq 0 \tag{4.11}$$

for any u in the subspace $\mathcal{D}(A)$. But,

$$\langle Au, u\rangle_{X_{\mathbf{K}}} = \sum_{i=1}^{n} \int_0^{d_i} \left(\frac{1}{r_i}\frac{d^2u_i}{dx^2} - g_i u_i\right)\overline{u_i}\,dx - \left\langle \tilde{G}_{21}\gamma_0 u^a + \tilde{G}_{22}u^b, u^b\right\rangle_{\mathbf{K}_r^m} .$$

If we integrate by parts and take into account that for $u \in \mathcal{D}(A)$,

$$\sum_{i=1}^{n} \frac{1}{r_i}\bigg|_0^{d_i} \frac{du_i}{dx}\overline{u_i} = \langle\gamma_1 u^a, \gamma_0 u^a\rangle_{\mathbf{K}^{2n}} = -\left\langle \hat{G}_{11}\gamma_0 u^a + \hat{G}_{12}u^b, \gamma_0 u^a\right\rangle_{\mathbf{K}^{2n}} ,$$

we obtain

$$\begin{aligned}
\langle Au, u\rangle_{X_{\mathbf{K}}} = &-\sum_{i=1}^{n} \frac{1}{r_i}\int_0^{d_i} \left|\frac{du_i}{dx}\right|^2 dx - \left\langle \hat{G}_{11}\gamma_0 u^a + \hat{G}_{12}u^b, \gamma_0 u^a\right\rangle_{\mathbf{K}^{2n}} \\
&-\sum_{i=1}^{n} g_i \int_0^{d_i} |u_i|^2\,dx - \left\langle \hat{G}_{21}\gamma_0 u^a + \hat{G}_{22}u^b, u^b\right\rangle_{\mathbf{K}^m} .
\end{aligned} \tag{4.12}$$

Therefore

$$\mathrm{Re}\langle Au, u\rangle_{X_{\mathbf{K}}} \leq -\sum_{i=1}^{n} g_i \int_0^{d_i} |u_i|^2\,dx - \mathrm{Re}\left\langle G\begin{bmatrix}\gamma_0 u^a \\ u^b\end{bmatrix}, \begin{bmatrix}\gamma_0 u^a \\ u^b\end{bmatrix}\right\rangle_{\mathbf{K}^{2n+m}} \tag{4.13}$$

and (4.11) is proved.

Also from (4.13) it follows that if $G > 0$ and $u \in \mathcal{D}(A), u \neq 0$, we have $\mathrm{Re}\langle Au, u\rangle_{X_{\mathbf{K}}} < 0$ i.e. strong dissipativity. $\qquad\square$

Lemma 4.2.

 a) If $G = G^{tr} \geq 0$ and $\mathbf{K} = \mathbf{C}$, then A is "m"–dissipative on $X_{\mathbf{K}}$.

 b) If $G \geq 0$ and $\mathbf{K} = \mathbf{R}$, then A is "m"–dissipative on $X_{\mathbf{K}}$.

Proof. We have to prove that for all $\lambda > 0$, $\mathcal{R}(\lambda \mathcal{J} - A) = X_{\mathbf{K}}$ i.e. for $f = \begin{bmatrix} f^a \\ f^b \end{bmatrix} \in X_{\mathbf{K}}$ there exists $u = \begin{bmatrix} u^a \\ u^b \end{bmatrix} \in \mathcal{D}(A)$ such that

$$-N \frac{d^2 u^a}{dx^2} + (P + \lambda \mathcal{I}^a) u^a = f^a \tag{4.14}$$

and

$$\tilde{G}_{21} \gamma_0 u^a + \tilde{G}_{22} u^b + \lambda u^b = f^b . \tag{4.15}$$

Above, $\mathcal{I}$ is the identity operator in $X_{\mathbf{K}}$ and $\mathcal{I}^a$ is the nth order unity matrix.

Let us take the space $Y_{\mathbf{K}} = H_1^n(\mathbf{K}) \times \mathbf{K}_s^m$ with the norm $\|h\|_Y^2 = \|h^a\|_{H_1^n}^2 + \|h^b\|_{\mathbf{K}_s^m}^2$. For $u, v \in Y_{\mathbf{K}}$ let us denote

$$a_1(u,v) = \left\langle N \frac{du^a}{dx}, \frac{dv^a}{dx} \right\rangle_{L_2^n}$$

$$a_2(u,v) = \langle (P + \lambda \mathcal{I}^a) u^a, v^a \rangle_{L_2^n}$$

$$a_3(u,v) = \left\langle \begin{bmatrix} \widehat{G}_{11} & \widehat{G}_{12} \\ \widehat{G}_{21} & \lambda \mathrm{diags}_i + \widehat{G}_{22} \end{bmatrix} \begin{bmatrix} \gamma_0 u^a \\ u^b \end{bmatrix}, \begin{bmatrix} \gamma_0 v^a \\ v^b \end{bmatrix} \right\rangle_{\mathbf{K}^{2n+m}} .$$

Let $a(\cdot,\cdot) : Y_{\mathbf{K}} \times Y_{\mathbf{K}} \to \mathbf{K}$ be the sesquilinear form

$$a(u,v) = a_1(u,v) + a_2(u,v) + a_3(u,v).$$

On the other hand, it is well known (cf. Agmon [1965]) that on H_1^n the following norms are equivalent

$$\|w\|_{H_1^n}^2 = \|w\|_{L_2^n}^2 + \left\| \frac{dw}{dx} \right\|_{L_2^n}^2 \quad \text{and}$$

$$\|w\|_{H_1^n}^2 = \|(w_1(x_1), \ldots, w_p(x_p))^{tr}\|_{\mathbf{K}^p}^2 + \|(w_{p+1}, \ldots, w_n)^{tr}\|_{L_2^{n-p}}^2 + \left\| \frac{dw}{dx} \right\|_{L_2^n}^2 ,$$

where $1 \leq p \leq n$ and $(x_1, \ldots, x_p)^{tr}$ is arbitrarily taken in $\prod_{i=1}^p [0, d_i]$.

Then, there clearly exist positive constants K_i such that

$$|a_i(u,v)| \leq K_i \|u\|_{Y_{\mathbf{K}}} \|v\|_{Y_{\mathbf{K}}} \quad \text{for } i = 1, 2, 3,$$

and this shows that the form $a(\cdot,\cdot)$ is bounded. On the other hand,

$$a_1(u,u) = \operatorname{Re} a_1(u,u) \geq \min_{1\leq i\leq n} \frac{1}{r_i c_i} \left\| \frac{du^a}{dx} \right\|_{L_2^n}^2$$

$$a_2(u,u) = \operatorname{Re} a_2(u,u) \geq \min_{1\leq i\leq n} \left(\frac{g_i}{c_i} + \lambda \right) \|u^a\|_{L_2^n}^2 \ .$$

At last, due to hypotheses, both in the cases $\mathbf{K} = \mathbf{C}$ and $\mathbf{K} = \mathbf{R}$ we have

$$a_3(u,u) = \operatorname{Re} a_3(u,u) = \left\langle G \begin{bmatrix} \gamma_0 u^a \\ u^b \end{bmatrix}, \begin{bmatrix} \gamma_0 u^a \\ u^b \end{bmatrix} \right\rangle_{\mathbf{K}^{2n+m}} + \lambda \left\langle (\operatorname{diag} s_i) u^b, u^b \right\rangle_{\mathbf{K}^m} \ . \tag{4.16}$$

Consequently,

$$a_3(u,u) \geq \lambda (\min_{1\leq i\leq n} s_i) \|u^b\|_{\mathbf{K}^m}^2 .$$

So $a(\cdot,\cdot)$ is a coercive form. Due to the Lax-Milgram lemma (see, for instance, Fattorini [1983 p.214]), there exists a unique $u \in Y_{\mathbf{K}}$ such that

$$a(u,w) = \langle f,w \rangle_{X_{\mathbf{K}}} \quad \text{for all } w \in Y_{\mathbf{K}} \ . \tag{4.17}$$

Particularly, we take $w = \begin{bmatrix} w^a \\ w^b \end{bmatrix}$ with $w^a \in \prod_{i=1}^{n} C_0^\infty(0,d_i;\mathbf{K})$ and $w^b = 0$. Because $\gamma_0 w^a = 0$, (4.17) implies

$$\sum_{i=1}^{n} c_i \int_0^{d_i} \left[-\frac{1}{r_i c_i} \frac{d^2 u_i}{dx^2} + \left(\frac{g_i}{c_i} + \lambda \right) u_i \right] \overline{w_i}\, dx = \sum_{i=1}^{n} c_i \int_0^{d_i} f_i \overline{w_i}\, dx \tag{4.18}$$

where $\dfrac{d^2 u_i}{dx^2}$ is the distributional derivative of $\dfrac{du_i}{dx} \in L_{2,i}$. It follows that, (see Yosida [1974 p.48])

$$-\frac{1}{r_i c_i} \frac{d^2 u_i}{dx^2}(x) + \left(\frac{g_i}{c_i} + \lambda \right) u_i(x) = f_i(x) \quad \text{a.e. in }]0,d_i[\quad \text{and for all } i = \overline{1,n}.$$

This implies $\dfrac{d^2 u_i}{dx^2} \in L_{2,i}$ and hence $u^a \in H_2^n$.

Then (4.18) can be written

$$\left\langle -N \frac{d^2 u^a}{dx^2} + (P + \lambda \mathcal{I}^a) u^a, w^a \right\rangle_{L_2^n} = \langle f^a, w^a \rangle_{L_2^n}$$

and because $\prod_{i=1}^{n} C_0^\infty(0,d_i;\mathbf{K})$ is dense in L_2^n we obtain (4.14).

Now we return to (4.17) and integrate by parts (see proof of Lemma 4.1), to obtain

$$\left\langle -N\frac{d^2u^a}{dx^2}, w^a \right\rangle_{L_2^n} + \langle \gamma_1 u^a, \gamma_0 w^a \rangle_{\mathbf{K}^{2n}} + a_2(u,v) + a_3(u,v) = \langle f, w \rangle_{X_{\mathbf{K}}} \quad (4.19)$$

for all $w \in Y_{\mathbf{K}}$. Taking $w^b = 0$ and using (4.14), this yields

$$\langle \gamma_1 u^a, \gamma_0 w^a \rangle_{\mathbf{K}^{2n}} + \left\langle \widehat{G}_{11}\gamma_0 u^a + \widehat{G}_{12}u^b, \gamma_0 w^a \right\rangle_{\mathbf{K}^{2n}} = 0$$

from which it follows

$$\gamma_1 u^a = -\widehat{G}_{11}\gamma_0 u^a - \widehat{G}_{12}u^b \quad (4.20)$$

i.e. $u \in \mathcal{D}(A)$. Finally, using (4.19) with (4.14) and (4.20) we find

$$\left\langle \widehat{G}_{21}\gamma_0 u^a + (\lambda \mathrm{diag} s_i + \widehat{G}_{22})u^b, w^b \right\rangle_{\mathbf{K}^m} = \langle f^b, w^b \rangle_{\mathbf{K}_s^m} \quad \text{for all } w^b \in \mathbf{K}_s^m.$$

Therefore (4.15) holds and this ends the proof of Lemma 4.2.   □

The above lemmas allow us to prove existence and uniqueness results.
Below, according to the similar definitions for the solutions of $(CP(\check{B}, u_0))$ used as in Section 1.4, we mean by <u>strong solution</u> of our dynamic problem $(P(B, v_0))$ a function in $C(0, T; X_{\mathbf{R}})$ which is absolutely continuous on each compact interval of $]0, T[$ and with components $v_1, v_2, \ldots, v_{n+m}$, satisfying $(P(B, v_0))$ for a.e. t and x. It is clear from this definition that time derivatives of the solution are considered in the $X_{\mathbf{R}}$ norm, while the space derivatives are of a distribution (generalized) type.

Also, by <u>generalized solution</u> of $(P(B, v_0))$ we mean a function $v \in C^1(0, T; X_{\mathbf{R}})$ satisfying $(P(B, v_0))$ for all t and a.e. x. Clearly a generalized solution is a strong solution.

Theorem 4.1.

a) *Suppose that there exists* $\nu \in]0, 1]$ *such that* $b_i \in C^{1+\nu}(0, T; \mathbf{R})$ *for* $i = \overline{1, n}$ *and* $b_i \in C^{\nu}(0, T; \mathbf{R})$ *for* $i = \overline{n+1, n+m}$, $G = G^{tr} \geq 0$ *and that* $v_0 \in X_{\mathbf{R}}$. *Then there exists a unique generalized solution of* $(P(B, v_0))$.

b) <u>*Suppose that*</u> $b_i \in C^2(0, T; \mathbf{R})$ *for* $i = \overline{1, n}$ *and* $b_i \in C^1(0, T; \mathbf{R})$ *for* $i = \overline{n+1, n+m}$, $b_i(0) = 0$, *for* $i = \overline{1, n}$, $G \geq 0$ *and that* $v_0 \in \mathcal{D}(A)$. *Then there exists a unique generalized solution of* $(P(B, v_0))$.

Proof. a) According to Lemma 4.1, Lemma 4.2 -part a) and according to the fact that on a Hilbert space a linear maximal dissipative operator is densely defined (see Brezis [1973,Prop.2.3]), the hypotheses of Lemma 1.13 are fulfilled (with $\delta = 0$). Then A generates an analytic C_0-contraction semigroup. On the other hand, the hypotheses imply (via (4.9) and (4.10)) $\tilde{B} \in C^{\nu}(0, T; X_{\mathbf{R}})$. Then Theorem 1.7 gives a unique classical solution for $(CP(\tilde{B}, u_0))$ with which (4.6), (4.7) and (4.9) assure the generalized solution v of $(P(B, v_0))$ belongs to the space $C^1(0, T; X_{\mathbf{R}})$.

b) Lemma 4.1 and part b) of Lemma 4.2 show that A is a linear, densely defined, "m"–dissipative operator on $X_{\mathbf{R}}$. Also $\tilde{B} \in C^1(0, T; X_{\mathbf{R}})$. From $b_i(0) = 0$ for $i = \overline{1, n}$ we deduce that $u^*(0) \in \mathcal{D}(A)$ and then $u_0 \in \mathcal{D}(A)$. Therefore, $(CP(\tilde{B}, u_0))$ has a unique classical solution (Theorem 1.4), and the result b) of Theorem 4.1 follows.   □

Let us try to discuss the significance of the above result. The hypotheses in the literature usually require very smooth sources (such as a constant or sinus functions). In our results, the initial conditions may be of "square-integrable type" (therefore, including discontinuous functions) and not even satisfying the boundary conditions (part a)). This allows us to include primitive models for transistors following specific goals, as we have discussed in Section 4.1. This is in accordance with the request $G = G^{tr}$ i.e. the network must not contain controlled sources. The main constraint is the semipositivity of G. If we use improved models for transistors assuring that the initial conditions satisfy the boundary ones, then naturally we have to accept non-reciprocal networks ($G \neq G^{tr}$). With a little price paid ($b_i(0) = 0$), the constraints for the model validation are the same, $G \geq 0$.

On the other hand, let us note that, if we consider smoother initial conditions, from the well-known a-priori estimates for parabolic problems combined with Sobolev's imbedding theorem (see for example Lions and Magenes [1972] or, for a more classical approach Hellwig [1967]), we can obtain a solution in the classical sense, i.e. with differentiability properties commonly used. In fact we have –Marinov, Neittaanmäki [1988]:

Theorem 4.2. *Let us consider all independent sources having simultaneous step variation at $t = 0$. Then, if $G \geq 0$ and $v_{k,0} \in C^2(0, d_k; \mathbf{R})$, $k = \overline{1, n}$ together with $v_{n+k,0}$, $k = \overline{1, m}$ satisfy the boundary conditions (SBC), then the problem $(P(B, v_0))$ has a unique solution in classical sense.*

Referring to the first example above (Figures 4.2 and 4.3) we find

$$
G = \begin{bmatrix}
G_1 + G_2 & & & & & \\
 & G_3 + G_6 & -G_3 & & & -G_6 \\
 & -G_3 & G_3 & & & \\
 & & & G_4 & -G_4 & \\
 & & & -G_4 & G_4 + G_5 & \\
 & -G_6 & & & & G_6 + G_7
\end{bmatrix}
$$

and $B = (V_d G_2, 0, ..., 0)^{tr}$, where we have denoted $G_i = 1/R_i$. Because $G = G^{tr} > 0$, the conclusion of Theorem 4.2 holds for $V_D =$ constant and any initial condition in $X_{\mathbf{R}}$.

For the second example (Figure 4.4),

$$G = \begin{bmatrix} 0 & 0 & 0 \\ 0 & 1/R & -1/R \\ 0 & -1/R & 1/R \end{bmatrix} \quad \text{and } B = (J, 0, 0)^{tr} .$$

Clearly $G = G^{tr} \geq 0$ and that is why the generalized solution of the dynamic process exists for a source $J \in C^{1+\nu}(0, T; \mathbf{R})$ and for square integrable initial conditions.

Let us consider now the usual case in digital circuits which have non-smooth sources, such as square pulses (i.e. sequences of Heaviside functions) or trapezoid pulses (i.e. differentiable functions with discontinuous derivatives). In this case we shall not expect a solution in the above sense. However, we can find a sequence of associate problems with $(P(B, v_0))$ which "converges" to $(P(B, v_0))$ and has a convergent sequence of strong solutions. The limit of this sequence of solutions will be called a "weak solution" of our problem. More precisely, a function $v \in L_1(0, T; X_{\mathbf{R}})$ is a <u>weak solution</u> of $(P(B, v_0))$ if there exist sequences $\{B^i\}_i$ and $\{v_0^i\}_i$ with the properties

- $\{B^i\}_i \to B$ in $L_1(0, T; \mathbf{R}^{2n+m})$ and $\{v_0^i\}_i \to v_0$ in $X_{\mathbf{R}}$.
- For each i the problem $(P(B^i, v_0^i))$ has a unique strong solution v^i.
- $\{v^i\}_i \to v$ in $L_1(0, T; X_{\mathbf{R}})$.

Theorem 4.3. If $B \in L_1(0, T; \mathbf{R}^{2n+m})$, $G \geq 0$ and $v_0 \in X_{\mathbf{R}}$, then $(P(B, v_0))$ has a unique weak solution $v \in L_1(0, T; X_{\mathbf{R}})$.

Due to this theorem, both our circuit examples have unique weak solutions if they are excited with discontinuous inputs such as the usual sequences of pulses. Again, the semipositivity of G is essential.

Proof. Let $u_0 = v_0 - u^*(0)$ where $u^*(0)$ is given by (4.7) and (4.9). It results in $u_0 \in X_{\mathbf{R}}$ and because $\overline{\mathcal{D}(A)} = X_{\mathbf{R}}$, for all i there exist $u_0^i \in \mathcal{D}(A)$ such that $\{u_0^i\}_i \to u_0$ in $X_{\mathbf{R}}$. We shall choose $v_0^i = u_0^i + u^*(0)$. On the other hand, let B^i be step functions such that $\{B^i\}_i \to B$ in $L_1(0, T; \mathbf{R}^{2n+m})$ and $B^i(0) = B(0)$ for all i. Let us consider the problem $(CP(\tilde{B}^i, u_0^i))$ where $\tilde{B}^i$ is a step function obtained through (4.9) and (4.10) and by extension in the non-differentiable points of B^i. Let $0 = a_0 < a_1 < \ldots < a_n = T$ be the partition of $[0, T]$ such that $\tilde{B}^i(t) \equiv z_k$ on $[a_{k-1}, a_k[$. If we denote by $S_k(t)$ the semigroup generated by the maximal dissipative operator $A + z_k \mathcal{J}$, and we define $u^i(t)$ by $u^i(0) = u_0^i$ and $u^i(t) = S_k(t - a_{k-1})u(a_{k-1})$ for $t \in [a_{k-1}, a_k]$, it is clear that u^i is a unique strong solution of $(CP(\tilde{B}^i, u_0^i))$. From Lemma 1.15 and Lemma 4.1 we find:

$$\frac{d}{dt}\|u^i(t) - u^j(t)\|_{X_{\mathbf{R}}} \leq \|\tilde{B}^i(t) - \tilde{B}^j(t)\|_{X_{\mathbf{R}}} \quad \text{a.e. in } [0, T] .$$

Consequently,

$$\|u^i(t) - u^j(t)\|_{X_\blacksquare} \le \|u^i(0) - u^j(0)\|_{X_\blacksquare} + \int_0^t \|\tilde{B}^i(s) - \tilde{B}^j(s)\|_{X_\blacksquare} \, ds$$

for all $i, j \in \overline{1, n + m}$ and $t \in [0, T]$.

This shows that $\{u^i(t)\}_i \to u(t)$ uniformly in $X_{\mathbf{R}}$ and then $u \in C(0, T; X_{\mathbf{R}})$. But by (4.6) $v^i = u^i + u^{i*}$ is the strong solution of $(P(B^i, v_0^i))$ and in addition $\{u^{i*}\}_i \to u^*$ in $L_1(0, T; X_{\mathbf{R}})$. Therefore $\{v^i\}_i \to u + u^* = v$ in $L_1(0, T; X_{\mathbf{R}})$. This $v \in L_1(0, T; X_{\mathbf{R}})$ is the desired unique weak solution of $(P(B, v_0))$. $\square$

4.4. The steady state problem

If we suppose the existence of the inverse $\widehat{G}_{22}^{-1}$, the time independent problem $(SP(B)) = (SE) + (SBC)$ can be rewritten in the form

$$\frac{d^2 v_k}{dx^2} - g_k r_k v_k = 0, \quad x \in \,]0, d_k[, \; k = \overline{1, n} \tag{4.21}$$

$$\gamma_1 v^a = - \left[\widehat{G}_{11} - \widehat{G}_{12}\widehat{G}_{22}^{-1}\widehat{G}_{21} \right] \gamma_0 v^a - \widehat{G}_{12}\widehat{G}_{22}^{-1} B^b + B^a \tag{4.22}$$

$$v^b = \widehat{G}_{22}^{-1} \left(B^b - \widehat{G}_{21}\gamma_0 v^a \right) , \tag{4.23}$$

where $\gamma_0 v^a$ and $\gamma_1 v^a$ have the same meaning as in the preceding section and $B^a \in \mathbf{R}^{2n}$, $B^b \in \mathbf{R}^m$ are two vectors such that $B = \begin{bmatrix} B^a \\ B^b \end{bmatrix}$.

For the convience of future references let us list some hypotheses:

H1 : There exists at least an index i, $1 \le i \le n$, such that $g_i = 0$,
$\overline{\text{H1}}$: For all $i = \overline{1, n}$, $g_i > 0$,
H2 : $\widehat{G}_{22}$ is invertible
H3 : $\widehat{G}_{11} - \widehat{G}_{12}\widehat{G}_{22}^{-1}\widehat{G}_{21} > 0$,
$\overline{\text{H3}}$: $\widehat{G}_{11} - \widehat{G}_{12}\widehat{G}_{22}^{-1}\widehat{G}_{21} \ge 0$.

With these, we can announce the result:

Theorem 4.4.

a) *Let the assumptions* **H1** + **H2** + **H3** *or* **$\overline{\text{H1}}$** + **H2** + **$\overline{\text{H3}}$** *are valid. Then the problem* $(SP(B))$ *has a solution* $v \in \prod_{i=1}^{n} C^\infty(0, T; \mathbf{R}) \times \mathbf{R}^m$.

b) *If* $G > 0$ *and if* $(SP(B))$ *has a solution in a classical sense this solution is unique.*

Proof.

a) Let us take $g_k = 0$ for $k = \overline{1, k_o}$ and $g_k > 0$ for $k = \overline{k_o + 1, n}$. The following functions

$$\begin{cases} v_k(x) = M_k x + N_k, & k = \overline{1, k_o} \\ v_k(x) = M_k e^{a_k x} + N_k e^{-a_k x}, & k = \overline{k_o + 1, n} \end{cases} \qquad (4.24)$$

verify equations (4.21), where we have denoted $a_k = \sqrt{r_k g_k}$. Let us take also the vector $w \in \mathbf{R}^{2n}$ with components $w_{2k-1} = N_k$ and $w_{2k} = M_k d_k + N_k$ for $k = \overline{1, k_o}$ and also $w_{2k-1} = M_k + N_k$ and $w_{2k} = M_k e^{a_k d_k} + N_k e^{-a_k d_k}$ for $k = \overline{k_o + 1, n}$.

By checking the boundary conditions (4.22) for the functions (4.24) we obtain a linear system of equations in $\mathbf{R}^{2n}$:

$$Pw = -\overline{G}w + \overline{B}. \qquad (4.25)$$

Here P is a linear operator (matrix) in $\mathbf{R}^{2n}$ defined by

$$[Pw]_{2k-1} = -[Pw]_{2k} = \frac{1}{r_k} \frac{w_{2k-1} e^{-a_k d_k} - w_{2k-1}}{e^{a_k d_k} - e^{-a_k d_k}}$$

for $k = \overline{1, k_o}$ and

$$[Pw]_{2k-1} = \frac{a_k}{r_k} \frac{w_{2k-1} \sinh a_k d_k - w_{2k}}{\sinh a_k d_k}$$

$$[Pw]_{2k} = \frac{a_k}{r_k} \frac{w_{2k} \cosh a_k d_k - w_{2k-1}}{\sinh a_k d_k}$$

for $k = \overline{k_o + 1, n}$.

Furthermore, in (4.25) we have denoted

$$\overline{G} = \widehat{G}_{11} - \widehat{G}_{12} \widehat{G}_{22}^{-1} \widehat{G}_{21} \quad \text{and}$$
$$\overline{B} = -\widehat{G}_{12} \widehat{G}_{22}^{-1} B^b + B^a .$$

The following estimate is easily derived for the operator P:

$$\langle Pw, w \rangle_{\mathbf{R}^{2n}} \geq M_1 \sum_{k=1}^{k_0} (w_{2k-1} - w_{2k})^2 + M_2 \sum_{k=k_0+1}^{n} (w_{2k-1}^2 + w_{2k}^2)$$

for all $w \in \mathbf{R}^{2n}$ and where M_1 and M_2 are strictly positive constants. Hence, the matrix $P + \overline{G}$ is positive definite under each of the two sets of hypotheses. Consequently, the equation (4.25) has a solution which fixes the constants in (4.24). The existence is proved.

b) In the same way as in Section 4.3, the problem $(SP(B))$ can be written as a problem on the space $X_\mathbf{R}$, namely

$$0 = Au + \tilde{B} , \tag{4.26}$$

where A and $\tilde{B}$ have the same definition as quoted in Section 4.3, but $\alpha_k, \beta_k, \delta_k$ are independent of time. Of course, for the present context, the constants c_k and s_k are artificially introduced to keep the previous notations.

If $\overline{u} \in \mathcal{D}(A)$ is another solution of $(SP(B))$ (and, hence of (4.26)) then,

$$\langle A(u - \overline{u}), u - \overline{u} \rangle_{X_\mathbf{R}} = \langle \tilde{B} - \tilde{B}, u - \overline{u} \rangle = 0. \tag{4.27}$$

On the other hand, if $G > 0$ then, by Lemma 4.1, A is strongly dissipative. This means that, supposing $\overline{u} \neq u$, we have

$$\langle A(u - \overline{u}), u - \overline{u} \rangle_{X_\mathbf{R}} < 0$$

which contradicts (4.27). Therefore $\overline{u} = u$.   □

The fact that $G > 0$ implies $\widehat{G}_{22} > 0$ and $\widehat{G}_{11} - \widehat{G}_{12} \widehat{G}_{22}^{-1} \widehat{G}_{21} > 0$, combined with the above result gives:

Corollary 4.1. *If $g_i \geq 0$, $i = \overline{1, n}$ and $G > 0$ then $(SP(B))$ has a unique classical solution.*

If we try to apply these results to our above examples of Section 4.1 we find

- for the first example, where $G > 0$, we have a unique steady state solution even if $g_i = 0$ for $i = 1$ or/and $i = 2$,
- for the second example, $G \geq 0$, $\widehat{G}_{22}^{-1}$ exists and $\widehat{G}_{11} - \widehat{G}_{12} \widehat{G}_{22}^{-1} G_{21} = 0$. The above results assure the existence of direct current solution only for $g \neq 0$ and does not affirm anything in the case $g = 0$. Because we know that in this last case the solution does not exist (see Section 4.1), we conclude that our sufficient conditions for existence are very close to the necessary ones.

4.5. Other qualitative results

A good model of a real process is one whose accuracy can be improved at will by more and more precise measurements of the inputs and parameters. In this respect, the following theorem will give sufficient conditions for the continuous dependence of the solution upon the data, the sources (vector B) and the initial conditions (vector v_0).

Theorem 4.5. *Let $G \geq 0$ and let $\{B^i\}_i$ be the sequences with elements in $L_1(0,T;\mathbf{R}^{2n+m})$ and $\{v_0^i\}_i$ a sequence with elements from $X_{\mathbf{R}}$ such that $\{B^i\}_i \to B$ in $L_1(0,T;\mathbf{R}^{2n+m})$ and $\{v_0^i\}_i \to v_0$ in $X_{\mathbf{R}}$ when $i \to \infty$. If v and v^i are the weak solutions for the problems $(P(B,v_0))$ and $(P(B^i,v_0^i))$ respectively, then $\{v^i\} \to v$ in $L_1(0,T;X_{\mathbf{R}})$.*

Proof. We may consider B^i and B as step functions. Otherwise, the assertion can be obtained by passing to the limit the result with step varying functions.

Let u and u^i be weak solutions (see Brezis [1973 p.64]) for the problems $(CP(\tilde{B}^i,u_0^i))$ and $(CP(\tilde{B},u_0))$ respectively (obtained by using (4.6), (4.7), (4.9) and (4.10)). The dissipativity of A combined with Lemma 1.15 gives

$$\|u^i(t) - u(t)\|_{X_{\mathbf{R}}} \leq \|u^i(0) - u(0)\|_{X_{\mathbf{R}}} + \int_0^t \|\tilde{B}^i(s) - \tilde{B}(s)\|_{X_{\mathbf{R}}} \, ds \ .$$

Using the fact that $\{u^{i*}(0)\}_i \to u^*(0)$ in $\mathbf{R}^{2n+m}$ and $\{u^{i*}\}_i \to u^*$ in $L_1(0,T;X_{\mathbf{R}})$ we obtain the result of Theorem 4.5. $\square$

At last, a property which the digital circuits might have is that if the sources are constant functions suddenly connected then the outputs must tend to constant values. As the following theorem shows, this property is obtained by strengthening the hypotheses.

Theorem 4.6. *Let us consider $g_i > 0$ for all $i = \overline{1,n}$, $B(t) = B$ is constant for $t \geq 0$, $v_0 \in X_{\mathbf{R}}$ and suppose there exists $\alpha > 0$ such that*

$$\langle Gx, x \rangle_{\mathbf{R}^{2n+m}} \geq \alpha \sum_{i=1}^m x_{2n+i}^2 \quad \text{for all } x \in \mathbf{R}^{2n+m}.$$

If v is a strong solution of $(P(B,v_0))$ and v_∞ is a solution (in a classical sense) for $(SP(B))$, then for all $t \geq 0$ we have

$$\|v(t) - v_\infty\|_{X_{\mathbf{R}}} \leq \|v_0 - v_\infty\|_{X_{\mathbf{R}}} e^{-\beta t}$$

where $\beta = \min(g_1, \ldots, g_n, \alpha)$.

Proof. As we have mentioned in the proof of Theorem 4.4, $(SP(B))$ can be written as $Au_\infty + \tilde{B} = 0$ where $u_\infty = v_\infty - u^*$ and $u^*, \tilde{B}$ are given by (4.7) and (4.10) with time independent terms.

But this problem can be seen as a Cauchy one, namely

$$\begin{cases} \dfrac{du_\infty}{dt} = Au_\infty + \tilde{B} \\ u_\infty(0) = u_\infty \ . \end{cases}$$

On the other hand, $(P(B, v_0))$ is associated with the problem $(CP(\tilde{B}, u_0))$ with the strong solution $u = v - u^*$. Thus, we obtain

$$\begin{cases} \dfrac{d}{dt}(u - u_\infty) = A(u - u_\infty) & \text{a.e. in } [0, T] \\ (u - u_\infty)(0) = u - u_\infty \in X_{\mathbf{R}} \ . \end{cases} \tag{4.28}$$

On the other hand, the hypothesis about the matrix G implies (see relation (4.13) in the proof of Lemma 4.1)

$$\langle Aw, w\rangle_{X_\blacksquare} \leq -\beta\|w\|^2_{X_\blacksquare} \quad \text{for all } w \in \mathcal{D}(A).$$

This fact, Lemma 1.15 and (4.28) give

$$\frac{d}{dt}\|u - u_\infty\|_{X_\blacksquare} \leq -\beta\|u - u_\infty\|_{X_\blacksquare}$$

a.e. in $[0, T]$ and for $u(t) \neq u_\infty$. From here we obtain the desired inequality. $\square$

The above result is an asymptotic stability property (of the global exponential type) of the d.c. solution: regardless of the initial conditions, the solution tends to the same steady state value.

Our first circuit example satisfies the conditions of the theorem with $\alpha = \min(G_5, G_7)$, while for the second circuit from Fig 4.4, (even with $g > 0$, when the steady state solution exists) the asymptotic convergence is not warranted by the above result.

4.6. Bibliographical comments

The results presented in this chapter were obtained in Marinov and Lehtonen [1989] and Marinov and Neittaanmäki [1988]. A different approach leading to a variational solution can be found in Marinov and Moroşanu [1991]. On the other hand, because in practice the nonlinearity of pull-up transistor in a MOS driver significantly influences the delay time (Wyatt [1985]), a similar problem to the above one but containing a nonlinear resistive part (i.e. nonlinear boundary conditions) is also interesting: Moroşanu, Marinov and Neittaanmäki [1989,1991]. Some results on nonlinear parabolic systems with very general nonlinear boundary conditions can be found in Moroşanu and Petrovanu [1986] and Moroşanu [1988].

The reader has probably remarked that the mathematical interest in our problem $(E) + (BC) + (IC)$ lies on very special boundary conditions. On the one hand they are of "crossed type", i.e. the value of a derivative at a boundary point depends on the value of the function at all boundary points. On the other hand, boundary conditions contain time derivatives. The fact that this type of boundary condition appears in transmission line problems was observed a long time ago: Brayton and Miranker [1964], Cooke and Krumme [1968]. They refer to the complete Telegraph Equations, which are of the hyperbolic type. Other qualitative studies on nonlinear hyperbolic equations with nonlinear boundary conditions (of crossed type and even with time derivatives) are: Barbu [1977], Barbu and Moroşanu [1981], Moroşanu [1981a,b,1982,1988]. Although above we consider degenerate Telegraph Equations (neglecting the inductance), our results can not be derived by those referring to the hyperbolic case.

Other recent studies regarding the correctness of distributed parameter models for integrated circuits are Showalter and Snyder [1986], Bose and Showalter [1990], Showalter and Xu [1990].

Chapter V

Asymptotic behaviour of mixed-type circuits.
Delay time predicting

5.0. Introduction

In the preceding chapter we have shown that the delay time problem in integrated circuits leads us to consider mixed-type circuits with distributed elements described by Telegraph Equations and lumped resistive and capacitive elements (Figure 4.5). Moreover, the well-posedness of the mathematical model $(P(B, v_0)) = (E) + (BC) + (IC)$ has been studied, various conditions for the existence, uniqueness and L_2-stability of different kind of solutions being formulated.

Let us consider now an integrated circuit whose delay time (especially caused by interconnections) we want to evaluate. We assume that the corresponding mixed-type circuit has unique dynamic and steady state solutions in a classical sense. (Sufficient conditions for this are given in Theorems 4.2 and 4.4.) The first problem solved below (in Section 5.2) is to choose precise supplementary conditions in which, irrespective of initial conditions, the dynamic solution tends to the steady state (this is the so called "global asymptotic stability" of the steady state). The convergence here is in the space of continuous functions of t and x, and the transient regime is provoked by a step variation of a part of the sources, the other sources remaining constant. It is exactly the interesting case in the study of the switching speed in integrated structures (see Marinov and Neittaanmäki [1989] and the examples in Sections 4.1 and 6.6). In Section 5.1 we define "the global delay time" as a performance parameter which expresses the rate of evolution of the whole network from initial conditions towards steady state. Because the stability theorem found here assures even the exponential type of stability, it is possible to infer an upper bound of the global delay time. As many examples show (see Section 6.6) this upper bound is sufficiently tight and can be itself considered as a global delay time. It is a very attractive parameter for circuit designers due to its closed form and the simplicity of the calculus. The inclusion of this delay time in a CAD (Computer

Aided Design) timing analyzer is perfectly possible. Let us remark that, from the mathematical point of view, the dissipativity of the abstract operator governing our problem (in the space of continuous functions) again plays the main part in our reasoning. The fact that the diagonal dominance of the matrix G is a condition for our results to work, does not significantly reduce the area of their applicability. In this respect the examples given in Section 6.7 are relevant.

A nonlinear case in which lumped nonlinearly modelled bipolar transistors are interconnected by rcg-lines is treated in Section 5.3.

The framing of our study in abundant engineering and mathematical literature devoted to the stability and delay time problem is analysed in Section 5.4.

5.1. Remarks on delay time evaluation

We deal with the dynamic process of the general network from Figure 4.5, after switching the constant sources at $t = 0$. So we have the problem $(P(B, v_0)) = (E) + (BC) + (IC)$ from the previous chapter, which we rewrite here:

(E)
$$\frac{\partial v_k}{\partial t} = \frac{1}{r_k c_k} \frac{\partial^2 v_k}{\partial x^2} - \frac{g_k}{c_k} v_k, \quad t \geq 0, \ x \in]0, d_k[, \ k = \overline{1, n}$$

with boundary conditions

(BC)
$$\begin{bmatrix} -\dfrac{1}{r_1}\dfrac{\partial v_1}{\partial x}(t, 0) \\[1.5ex] +\dfrac{1}{r_1}\dfrac{\partial v_1}{\partial x}(t, d_1) \\[1ex] \vdots \\[1ex] -\dfrac{1}{r_n}\dfrac{\partial v_n}{\partial x}(t, 0) \\[1.5ex] +\dfrac{1}{r_n}\dfrac{\partial v_n}{\partial x}(t, d_n) \\[1.5ex] s_1\dfrac{dv_{n+1}}{dt}(t) \\[1.5ex] \vdots \\[1ex] s_m\dfrac{dv_{n+m}}{dt}(t) \end{bmatrix} = -G \begin{bmatrix} v_1(t, 0) \\[1.5ex] v_1(t, d_1) \\[1ex] \vdots \\[1ex] v_n(t, 0) \\[1.5ex] v_n(t, d_n) \\[1.5ex] v_{n+1}(t) \\[1.5ex] \vdots \\[1ex] v_{n+m}(t) \end{bmatrix} + B, \quad t \geq 0,$$

and with initial conditions

(IC)
$$\begin{cases} v_k(0, x) = v_{k,0}(x), & x \in]0, d_k[, \ k = \overline{1, n} \\ v_{n+k}(0) = v_{n+k,0}, & k = \overline{1, m}. \end{cases}$$

Above, B is a constant vector obtained from the constant value of sources. The steady state to which the transient regime tends, is described by

$$(SP(B)) = (SE) + (SBC):$$

(SE) $\qquad \dfrac{d^2 \tilde{v}_k}{dx^2} - g_k r_k \tilde{v}_k = 0, \quad x \in]0, d_k[, \; k = \overline{1, n}$

(SBC)
$$
\begin{bmatrix}
-\dfrac{1}{r_1}\dfrac{d\tilde{v}_1}{dx}(0) \\[2mm]
+\dfrac{1}{r_1}\dfrac{d\tilde{v}_1}{dx}(d_1) \\[1mm]
\vdots \\[1mm]
-\dfrac{1}{r_n}\dfrac{d\tilde{v}_n}{dx}(0) \\[2mm]
+\dfrac{1}{r_n}\dfrac{d\tilde{v}_n}{dx}(d_n) \\[1mm]
0 \\[1mm]
\vdots \\[1mm]
0
\end{bmatrix}
= -G
\begin{bmatrix}
\tilde{v}_1(0) \\[1mm]
\tilde{v}_1(d_1) \\[1mm]
\vdots \\[1mm]
\tilde{v}_n(0) \\[1mm]
\tilde{v}_n(d_n) \\[1mm]
\tilde{v}_{n+1} \\[1mm]
\vdots \\[1mm]
\tilde{v}_{n+m}
\end{bmatrix}
+ B \; .
$$

To facilitate the writing below, we shall consider even the capacitor voltages as space dependent functions:

$v_k : [0, \infty[\times [0, d_k] \to \mathbf{R}$ where $d_k = 0$ for $k = \overline{n+1, n+m}$ in $(P(B, v_0))$ and

$\tilde{v}_k : [0, d_k] \to \mathbf{R}$ where also $d_k = 0$ for $k = \overline{n+1, n+m}$ in $(SP(B))$.

We shall suppose throughout this chapter that the problem $(P(B, v_0))$ as well as $(SP(B))$ has a unique solution in the classical sense. We have denoted these solutions by $v = (v_1, \ldots, v_{n+m})$ and by $\tilde{v} = (\tilde{v}_1, \ldots, \tilde{v}_{n+m})$ respectively. Hence, our assumptions in what follows must be consistent with existence and uniqueness conditions (see Theorem 4.2 and Theorem 4.4).

The global dynamic behaviour of our network can be described by a function $D : [0, \infty[\to \mathbf{R}$, named "delay" and defined by

$$
D(t) = \frac{\displaystyle\max_{1 \leqslant i \leqslant n+m} \max_{0 \leqslant x_i \leqslant d_i} |v_i(t, x_i) - \tilde{v}_i(x_i)|}{\displaystyle\max_{1 \leqslant i \leqslant n+m} \max_{0 \leqslant x_i \leqslant d_i} |v_{i,0}(x_i) - \tilde{v}_i(x_i)|},
$$

supposing a-priori that $\tilde{v} \neq v_0 = (v_{1,0}, \ldots, v_{n+m,0})$.

When the asymptotic stability conditions are fulfilled, the delay variation begins from 1 (corresponding to the initial conditions) and tends to 0 (corresponding to the steady state) when time indefinitely grows. If we fix $\lambda \in (0,1)$, the speed of this evolution (and therefore the speed of signal propagation in our network) can be expressed by the last moment when the delay equals value λ. In this way, we are conducted to define "the (global) λ delay-time" as

$$T_\lambda = \sup\{t : D(t) = \lambda\} \ .$$

Below, we intend to give an upper bound of this parameter.

When we have a network of the form given in Figure 4.5, we arbitrarily denote the ends of the ith line by 0 and d_i , $i = \overline{1,n}$. The intuitive fact that the delay does not depend on line termination notation can be shown as follows. The first line terminals inversion is equivalent with a spatial variable change in $(P(B, v_0))$ and $(SP(B))$. If $y_1, \ldots y_{n+m}$ are the new variables and $x_1, \ldots, x_{n+m}$ the old ones, we have $y_1 = d_1 - x_1$, $y_2 = x_2$, $\ldots, y_{n+m} = x_{n+m}$. The new voltages for the dynamic problem are $u_1(t, y_1) = v_1(t, d_1 - y_1) = v_1(t, x_1)$, $u_2(t, y_2) = v_2(t, x_2), \ldots, u_{n+m}(t, y_{n+m}) = v_{n+m}(t, x_{n+m})$. Then we obtain the parabolic system of equations

$$(E^1) \qquad \frac{\partial u_k}{\partial t} = \frac{1}{r_k c_k} \frac{\partial^2 u_k}{\partial y^2} - \frac{g_k}{c_k} u_k, \quad t \geq 0, \ y \in]0, d_k[, \ k = \overline{1,n}$$

with boundary conditions

$$(BC^1) \qquad
\begin{bmatrix}
+\dfrac{1}{r_1}\dfrac{\partial u_1}{\partial y}(t, d_1) \\[2mm]
-\dfrac{1}{r_1}\dfrac{\partial u_1}{\partial y}(t, 0) \\[2mm]
-\dfrac{1}{r_2}\dfrac{\partial u_2}{\partial y}(t, 0) \\[2mm]
+\dfrac{1}{r_2}\dfrac{\partial u_2}{\partial y}(t, d_2) \\[2mm]
\vdots \\[2mm]
-\dfrac{1}{r_n}\dfrac{\partial u_n}{\partial y}(t, 0) \\[2mm]
+\dfrac{1}{r_n}\dfrac{\partial u_n}{\partial y}(t, d_n) \\[2mm]
s_1\dfrac{du_{n+1}}{dt} \\[2mm]
\vdots \\[2mm]
s_m\dfrac{du_{n+m}}{dt}
\end{bmatrix}
= -G
\begin{bmatrix}
u_1(t, d_1) \\[2mm]
u_1(t, 0) \\[2mm]
u_2(t, 0) \\[2mm]
u_2(t, d_2) \\[2mm]
\vdots \\[2mm]
u_n(t, 0) \\[2mm]
u_n(t, d_n) \\[2mm]
u_{n+1}(t) \\[2mm]
\vdots \\[2mm]
u_{n+m}(t)
\end{bmatrix}
+ B, \quad t \geq 0 \, ,$$

and with initial conditions

$$(IC^1) \qquad \begin{cases} u_k(0, y_k) = v_{k,0}(x_k), & y_k \in]0, d_k[, \ k = \overline{1,n} \\ u_{n+k}(0) = v_{n+k,0}, & k = \overline{1,m} \ . \end{cases}$$

Similarly, if we denote the new voltages in the direct current problem by $\tilde{u}_1(y_1) = \tilde{v}_1(d - y_1) = \tilde{v}_1(x_1)$, $\tilde{u}_2(y_2) = \tilde{v}_2(x_2)$, ... $\tilde{u}_{n+m}(y_{n+m}) = \tilde{v}_{n+m}(x_{n+m})$, we obtain new boundary conditions (SBC^1) which, confronted by (SBC), and have the same differences as (BC^1) confronted by (BC).

The change of variables we made, gives

$$\max_{y_k \in [0, d_k]} |u_k(t, y_k) - \tilde{u}_k(y_k)| = \max_{x_k \in [0, d_k]} |v_k(t, x_k) - \tilde{v}_k(x_k)|$$

for $k = \overline{1, n+m}$ and all $t > 0$. This shows that the delay (and consequently, the delay time) is invariant at line-terminals inversion.

Now, we shall try to put the problem $(P^1(B, v_0)) = (E^1) + (BC^1) + (IC^1)$ in a similar form to problem $(P(B, v_0))$. To this goal let us denote by $M \oplus N$ the matrix $\begin{bmatrix} M & 0 \\ 0 & N \end{bmatrix}$, where M and N are matrices even with different dimensions. Let I_k be the unity matrix with the dimension k, and for every $k = \overline{1, n}$ we denote

$$P_k^{\sigma_k} = \begin{cases} I_{2k-2} \oplus \begin{bmatrix} 0 & 1 \\ 1 & 0 \end{bmatrix} \oplus I_{2n+m-2k} & \text{when } \sigma_k = 1 \\ I_{2n+m} & \text{when } \sigma_k = 0 \ . \end{cases}$$

We can easily observe that, if R is a $2n + m$ vector, then $P_k^1 R$ is obtained from R by inverting the $(2k-1)$-th term with the $2k$-th one, while $P_k^0 R = R$. Also, if M is a $2n + m$ square matrix, $P_k^1 M P_k^1$ differs from M by interchanging the $(2k-1)$-th row with $2k$-th one and the $(2k-1)$-th column with $2k$-th one. Also $P_k^0 M P_k^0 = M$. Then, multiplying to the left the relations (BC^1) with P_1^1 we obtain

$$\begin{bmatrix} -\dfrac{1}{r_1} \dfrac{\partial u_1}{\partial y}(t, 0) \\ \dfrac{1}{r_1} \dfrac{\partial u_1}{\partial y}(t, d_1) \\ \vdots \\ -\dfrac{1}{r_n} \dfrac{\partial u_n}{\partial y}(t, 0) \\ \dfrac{1}{r_n} \dfrac{\partial u_n}{\partial y}(t, d_n) \\ s_1 \dfrac{du_{n+1}}{dt} \\ \vdots \\ s_m \dfrac{du_{n+m}}{dt} \end{bmatrix} = -P_1^1 G P_1^1 \begin{bmatrix} u_1(t, 0) \\ u_1(t, d_1) \\ \vdots \\ u_n(t, 0) \\ u_n(t, d_n) \\ u_{n+1}(t) \\ \vdots \\ u_{n+m}(t) \end{bmatrix} + P_1^1 B \ .$$

Therefore, the problem $(E^1) + (BC^1) + (IC^1)$ has the same form as the problem $(E) + (BC) + (IC)$, in which the matrix G is replaced by $P_1^1 G P_1^1$ and the vector B by $P_1^1 B$. Because the delay time is the same for the two problems, when we compute the upper bounds of this parameter (depending on G and B with respect to $P_1^1 G P_1^1$ and $P_1^1 B$) clearly we must take the minimum value to get the best upper bound.

Extending the above arguments, the delay time remains unchanged if in the problem $(E) + (BC) + (IC)$ we replace the matrix G with the matrix

$$G^\sigma = P_n^{\sigma_n} P_{n-1}^{\sigma_{n-1}} \cdots P_1^{\sigma_1} G P_1^{\sigma_1} \cdots P_{n-1}^{\sigma_{n-1}} P_n^{\sigma_n}$$

and the vector B with the vector

$$B^\sigma = P_n^{\sigma_n} P_{n-1}^{\sigma_{k-1}} \cdots P_1^{\sigma_1} B ,$$

where we have denoted $\sigma = (\sigma_1, \sigma_2, \ldots, \sigma_n)$. Depending on the values of σ, this new problem comprises of all possibilities for line-terminals inversion (for the above example $\sigma = (1, 0, 0, \ldots, 0)$). This is why throughout the following we shall consider our problem with the matrix G^σ (whose elements are G_{ij}^σ) and with the vector B^σ (with elements b_i^σ). For each σ we shall find an upper bound $\overline{T}_\lambda(\sigma)$ of the delay time, after which we shall minimize this value with respect to σ i.e. with respect to the 2^n possibilities appearing when we interchange the line-terminals. Therefore

$$\overline{T}_\lambda = \min_\sigma \overline{T}_\lambda(\sigma) . \tag{5.1}$$

5.2. Asymptotic stability. Upper bound of delay time

For reasons that will become clear below, in our problem $(E) + (BC) + (IC)$ with G^σ and B^σ, we shall make a change of functions

$$\begin{cases} v_k(t, x) = u_k(t, x) \cos \dfrac{\alpha_k^\sigma x}{d_k}, & k = \overline{1, n} \\[2mm] v_k(t, x) = u_k(t, x), & k = \overline{n+1, n+m} , \end{cases} \tag{5.2}$$

where α_k^σ will be conveniently chosen. Also, we shall extract the time derivatives from the boundary conditions and attach them to the system; so we shall obtain for $t \geq 0$

$$\begin{cases} \dfrac{\partial u_k}{\partial t} = \dfrac{1}{r_k c_k} \dfrac{\partial^2 u_k}{\partial x^2} - \left[\dfrac{2\alpha_k^\sigma}{r_k c_k d_k} \tan \dfrac{\alpha_k^\sigma x}{d_k} \right] \dfrac{\partial u_k}{\partial x} - \\[3mm] \qquad - \left[\dfrac{(\alpha_k^\sigma)^2}{r_k c_k d_k^2} + \dfrac{g_k}{c_k} \right] u_k, \quad x \in]0, d_k[, \; k = \overline{1, n} \\[3mm] \dfrac{\partial u_k}{\partial t} = -\dfrac{1}{s_{k-n}} \left[\displaystyle\sum_{j=n+1}^{n+m} G_{n+k,n+j}^\sigma u_j(t, 0) + \sum_{j=1}^{n} G_{n+k,2j-1}^\sigma u_j(t, 0) + \right. \\[3mm] \qquad \left. + \displaystyle\sum_{j=1}^{n} G_{n+k,2j}^\sigma u_j(t, d_j) \cos \alpha_j^\sigma - b_{n+k}^\sigma \right], \quad k = \overline{n+1, n+m} \end{cases} \tag{5.3}$$

and the remaining boundary conditions are:

$$
\begin{bmatrix}
-\dfrac{1}{r_1}\dfrac{\partial u_1(t,0)}{dx} \\[2mm]
\dfrac{1}{r_1}\dfrac{\partial u_1(t,d_1)}{\partial x}\cos\alpha_1^\sigma \\[2mm]
\vdots \\[2mm]
-\dfrac{1}{r_n}\dfrac{\partial u_n(t,0)}{\partial x} \\[2mm]
\dfrac{1}{r_n}\dfrac{\partial u_n(t,d_n)}{\partial x}\cos\alpha_n^\sigma
\end{bmatrix}
= -\overline{G}^\sigma
\begin{bmatrix}
u_1(t,0) \\
u_1(t,d_1)\cos\alpha_1^\sigma \\
\vdots \\
u_n(t,0) \\
u_n(t,d_n)\cos\alpha_n^\sigma \\
u_{n+1}(t,0) \\
\vdots \\
u_{n+m}(t,0)
\end{bmatrix}
+
\begin{bmatrix}
0 \\
\dfrac{\alpha_1 u_1(t,d_1)}{r_1 d_1}\sin\alpha_1^\sigma \\[2mm]
\vdots \\
0 \\
\dfrac{\alpha_n u_n(t,d_n)}{r_n d_n}\sin\alpha_n^\sigma
\end{bmatrix}
+\overline{B}^\sigma ,
$$

$$(5.4)$$

where $\overline{G}^\sigma$ and $\overline{B}^\sigma$ are formed with the first $2n$ rows from G^σ and B^σ respectively. The initial conditions become:

$$
\begin{cases}
u_k(0,x) = v_{k0}(x)/\cos\dfrac{\alpha_k^\sigma x}{d_k}, & x \in\,]0,d_k[,\ k=\overline{1,n} \\[3mm]
u_k(0,x) = v_{k0}(x), & x \in\,]0,d_k[,\ d_k=0,\ k=\overline{n+1,n+m} .
\end{cases}
$$

$$(5.5)$$

We denote $Y = \prod\limits_{i=1}^{n+m}[0,d_i]$ and define on the Banach space $C(Y;\mathbf{R}^{n+m})$ the subset:

$$
\mathcal{D}(A) = \Bigg\{ f \in C(Y;\mathbf{R}^{n+m});\;
\begin{bmatrix}
-\dfrac{1}{r_1}\dfrac{df_1}{dx}(0) \\[2mm]
\dfrac{1}{r_1}\dfrac{df_1}{dx}(d_1)\cos\alpha_1^\sigma \\[2mm]
\vdots \\[2mm]
-\dfrac{1}{r_n}\dfrac{df_n}{dx}(0) \\[2mm]
\dfrac{1}{r_n}\dfrac{df_n}{dx}(d_n)\cos\alpha_n^\sigma
\end{bmatrix}
= -\overline{G}^\sigma
\begin{bmatrix}
f_1(0) \\
f_1(d_1)\cos\alpha_1^\sigma \\
\vdots \\
f_n(0) \\
f_n(d_n)\cos\alpha_n^\sigma \\
f_{n+1}(0) \\
\vdots \\
f_{n+m}(0)
\end{bmatrix}
+
$$

$$
+
\begin{bmatrix}
0 \\
\dfrac{\alpha_1 f_1(d_1)}{r_1 d_1}\sin\alpha_1^\sigma \\[2mm]
\vdots \\
0 \\
\dfrac{\alpha_n f_n(d_n)}{r_n d_n}\sin\alpha_n^\sigma
\end{bmatrix}
+\overline{B}^\sigma
\quad
\begin{array}{l}
\text{and } f_1,\ldots,f_n \ \text{(the com-} \\
\text{ponents of } f) \text{ are twice} \\
\text{continuously differentiable}
\end{array}
\Bigg\}
$$

Also we define an operator $A : \mathcal{D}(A) \to C(Y; \mathbf{R}^{n+m})$ by

$$(Af)(x) = \begin{cases} \dfrac{1}{r_k c_k} \dfrac{d^2 f_k(x)}{dx^2} - \left(\dfrac{2}{r_k c_k} \dfrac{\alpha_k^\sigma}{d_k} \tan \dfrac{\alpha_k^\sigma x}{d_k} \right) \dfrac{df_k(x)}{dx} - \\ \qquad - \left[\dfrac{(\alpha_k^\sigma)^2}{d_k^2 r_k c_k} + \dfrac{g_k}{c_k} \right] f_k(x), \quad k = \overline{1, n} \\ -\dfrac{1}{s_{k-n}} \left[\sum_{j=n+1}^{n+m} G_{n+k,n+j}^\sigma f_j(0) + \sum_{j=1}^{n} G_{n+k,2j-1}^\sigma f_j(0) + \right. \\ \qquad \left. + \sum_{j=1}^{n} G_{n+k,2j}^\sigma f_j(d_j) \cos \alpha_j^\sigma - b_{n+k}^\sigma \right], \quad k = \overline{n+1, n+m} . \end{cases}$$

With these, our problem (5.3)+(5.4)+(5.5) is equivalent to an abstract Cauchy problem on the space $C(Y; \mathbf{R}^{n+m})$. Namely

$$\begin{cases} \dfrac{du}{dt} = Au \\ u(0, \cdot) = u_0 = \text{ a function with components given by (5.5).} \end{cases} \tag{5.6}$$

The following lemma is essential for deriving our result. According to (4.5) we shall denote $S_i^\sigma = \sum_{j=1, \ j \neq i}^{2n+m} |G_{ij}^\sigma|$.

Lemma 5.1. *Let us suppose that G is DRD (see (4.5)), and for every $j = \overline{1, n}$ let us consider $\gamma_j^\sigma \in]0, \pi/2[$ such that*

$$\cos \gamma_j^\sigma = \left[S_{2j}^\sigma + \sqrt{(S_{2j}^\sigma)^2 + \frac{4}{r_j d_j} \left(G_{2j,2j}^\sigma + \frac{1}{r_j d_j} \right)} \right] \Bigg/ 2 \left(G_{2j,2j}^\sigma + \frac{1}{r_j d_j} \right) . \tag{5.7}$$

If we choose $\alpha_j^\sigma = \gamma_j^\sigma - \epsilon$ where $\epsilon > 0$ is such that $\alpha_j^\sigma \in]0, \pi/2[$ and if

$$\omega_\epsilon^\sigma = \max \left\{ \max_{1 \leq j \leq n} \left[-\frac{(\gamma_j^\sigma - \epsilon)^2}{d_j^2 r_j c_j} - \frac{g_j}{c_j} \right]; \max_{n+1 \leq j \leq n+m} \left(\frac{-G_{n+j,n+j} + S_{n+j}}{s_{j-n}} \right) \right\} \tag{5.8}$$

then, the operator A is totally ω_ϵ^σ-dissipative.

Proof. By using Lemma 1.6 we have to prove that for any $f, \overline{f} \in \mathcal{D}(A)$, the following inequality holds:

$$\sup_{(j,x) \in M(w)} [Af - A\overline{f}]_j(x) \operatorname{sgn} w_j(x) \leq \omega_\epsilon^\sigma \|w\| , \tag{5.9}$$

where $w = f - \overline{f}$ and

$$M(w) = \left\{ (p, y) \mid \max_{1 \leqslant i \leqslant n+m} \max_{0 \leqslant x \leqslant d_i} |w_i(x)| = |w_p(y)| = \|w\| \right\}. \tag{5.10}$$

Let us consider, already proven, that if $(p, y) \in M(w)$ and $p = \overline{1, n}$, then $y \in]0, d_p[$.

If for $p = \overline{1, n}$ we suppose $w_p(y) \geqslant 0$, (5.10) shows that y is a maximum point of w_p in $]0, d_p[$, i.e. $\dfrac{dw_p}{dx}(y) = 0$ and $\dfrac{d^2 w_p}{dx^2}(y) \leq 0$ and then

$$[Af - A\overline{f}]_p(y)\,\mathrm{sgn}\,w_p(y)$$

$$= \frac{1}{r_p c_p} \frac{d^2 w_p}{dx^2}(y) - \left(\frac{2\alpha_p^\sigma}{r_p c_p d_p} \tan \frac{\alpha_p^\sigma y}{d_p} \right) \frac{dw_p}{dx}(y) - \left[\frac{(\alpha_p^\sigma)^2}{d_p^2 r_p c_p} + \frac{g_p}{c_p} \right] |w_p(y)|$$

$$\leq - \left[\frac{(\alpha_p^\sigma)^2}{d_p^2 r_p c_p} + \frac{g_p}{c_p} \right] \|w\| . \tag{5.11}$$

If for $p = \overline{1, n}$ we suppose $w_p(y) < 0$ where $(p, y) \in M(w)$, then y is a minimum point of w_p in $]0, d_p[$ and (5.11) is valid again, as we can easily observe.

Finally, if $(p, y) \in M(w)$ and $p = \overline{n+1, n+m}$ (that means $y = 0$) then $|w_p(0)| = |w_p(y)| \geqslant |w_i(x)|$ for any $i = \overline{1, n+m}$ and any $x \in [0, d_i]$. It follows

$$[Af - A\overline{f}]_p(y)\,\mathrm{sgn}\,w_p(y) = -\frac{\mathrm{sgn}\,w_p(y)}{s_{p-n}} \Bigg[G^\sigma_{n+p,n+p} w_p(y) +$$

$$+ \sum_{\substack{j=n+1 \\ j \neq p}}^{n+m} G^\sigma_{n+p,n+j} w_j(0) + \sum_{j=1}^{n} G^\sigma_{n+p,2j-1} w_j(0) + \sum_{j=1}^{n} G^\sigma_{n+p,2j} w_j(d_j) \cos \alpha_j^\sigma \Bigg]$$

and if we take into account that for $i = \overline{2n+1, 2n+m}$ $G_{ii}^\sigma = G_{ii}$ and $S_i^\sigma = S_i$ we obtain

$$[Af - A\overline{f}]_p(y)\,\mathrm{sgn}\,w_p(y) \leq \frac{-G_{n+p,n+p} + S_{n+p}}{s_{p-n}} \|w\| . \tag{5.12}$$

The inequalities (5.11) and (5.12) give us the desired result (5.9). It remains to prove that if $(p, y) \in M(w)$ and $p = \overline{1, n}$ then $y \neq 0$ and $y \neq d_p$. Let us suppose, by contradiction, that $p = \overline{1, n}$, $y = 0$ and $(p, y) \in M(w)$. Taking into account that $f, \overline{f} \in \mathcal{D}(A)$ and multiplying the odd rows in the definition of $\mathcal{D}(A)$ by $w_p(0)$, we obtain

$$-\frac{1}{r_p} \frac{dw_p(0)}{dx} w_p(0) = -\sum_{j=1}^{n} G^\sigma_{2p-1,2j-1} w_j(0) w_p(0) -$$

$$- \sum_{j=1}^{n} G^\sigma_{2p-1,2j} w_j(d_j) w_p(0) \cos \alpha_j^\sigma - \sum_{j=1}^{m} G^\sigma_{2p-1,2n+j} w_{n+j}(0) w_p(0) .$$

But, $(p, y) \in M(w)$ implies $|w_p(0)| \geqslant |w_i(x_i)|$ for all $i = \overline{1, n+m}$ and $x_i \in [0, d_i]$, and therefore

$$-\frac{1}{r_p}\frac{dw_p(0)}{dx}w_p(0) \leq -G_{2p-1,2p-1}^{\sigma}|w_p(0)|^2 + \sum_{\substack{j=1 \\ j \neq p}}^{n}|G_{2p-1,2j-1}^{\sigma}||w_p(0)|^2 +$$

$$+ \sum_{j=1}^{n}|G_{2p-1,2j}^{\sigma}||w_p(0)|^2 + \sum_{j=1}^{m}|G_{2p-1,2n+j}^{\sigma}||w_p(0)|^2$$

i.e.

$$-\frac{1}{r_p}\frac{dw_p(0)}{dx}w_p(0) \leq (-G_{2p-1,2p-1}^{\sigma} + S_{2p-1}^{\sigma})|w_p(0)|^2 .$$

But, as we can easily observe, the DRD property given for G, implies (in fact is equivalent to) the same property for G^{σ}. That is why the last inequality gives

$$-\frac{1}{r_p}\frac{dw_p(0)}{dx}w_p(0) < 0 . \tag{5.13}$$

On the other hand, if $w_p(0) \geqslant 0$, then $w_p(0) \geqslant w_i(x)$ for $i = \overline{1, n+m}$ and $x \in [0, d_i]$, such that

$$\left[-\frac{1}{r_p}\lim_{x \to 0+}\frac{w_p(x) - w_p(0)}{x}\right]w_p(0) \geq 0 . \tag{5.14}$$

If $w_p(0) < 0$, then $w_p(0) \leqslant w_i(x)$ which again implies (5.14). But (5.14) contradicts (5.13) and this means that the initial assumption is false, i.e. $y \neq 0$.

Now, let us suppose that $p = \overline{1, n}$, $y = d_p$ and $(p, y) \in M(w)$. By utilizing even rows in the definition of $\mathcal{D}(A)$ we find as above:

$$\frac{1}{r_p}\frac{dw_p(d_p)}{dx}w_p(d_p)\cos\alpha_p^{\sigma}$$

$$\leq \left(-G_{2p,2p}^{\sigma}\cos\alpha_p^{\sigma} + \sum_{\substack{j=1 \\ j \neq 2p}}^{2n}|G_{2p,j}^{\sigma}| + \sum_{j=1}^{m}|G_{2p,2n+j}^{\sigma}| + \frac{\alpha_p^{\sigma}}{r_p d_p}\sin\alpha_p^{\sigma}\right)|w_p(d_p)|^2$$

$$= \left(-G_{2p,2p}^{\sigma}\cos\alpha_p^{\sigma} + S_{2p}^{\sigma} + \frac{\alpha_p^{\sigma}}{r_p d_p}\sin\alpha_p^{\sigma}\right)|w_p(d_p)|^2 . \tag{5.15}$$

On the other hand, from (5.7) we obtain

$$\cos\alpha_p^{\sigma} > \left[S_{2p}^{\sigma} + \sqrt{(S_{2p}^{\sigma})^2 + \frac{4}{r_p d_p}(G_{2p,2p}^{\sigma} + \frac{1}{r_p d_p})}\right] \bigg/ 2(G_{2p,2p}^{\sigma} + \frac{1}{r_p d_p})$$

which implies

$$-(G^\sigma_{2p,2p} + \frac{1}{r_p d_p}) \cos^2 \alpha^\sigma_p + S^\sigma_{2p} \cos \alpha^\sigma_p + \frac{1}{r_p d_p} < 0 \ .$$

From here, with inequality $\alpha^\sigma_p < \tan \alpha^\sigma_p$ valid for $\alpha^\sigma_p \in]0, \frac{\pi}{2}[$ we derive

$$-G^\sigma_{2p,2p} \cos \alpha^\sigma_p + S^\sigma_{2p} + \frac{\alpha^\sigma_p}{r_p d_p} \sin \alpha^\sigma_p < 0 \ .$$

Consequently, (5.15) yields

$$\frac{1}{r_p} \frac{dw_p(d_p)}{dx} w_p(d_p) \cos \alpha^\sigma_p < 0 \ . \tag{5.16}$$

On the other hand, if $w_p(d_p) \geqslant 0$, then $w_p(d_p) \geqslant w_i(x)$ for any $i = \overline{1, n+m}$ and $x \in [0, d_i]$. Thus,

$$\frac{1}{r_p} \lim_{x \to o^+} \frac{w_p(d_p) - w_p(d_p - x)}{x} w_p(d_p) \cos \alpha^\sigma_p \geq 0 \ . \tag{5.17}$$

The same inequality can be obtained if $w_p(d_p) < 0$.

Since (5.16) and (5.17) are contradictory, we conclude that for $p = \overline{1, n}$ and $(p, y) \in M(w)$ we have $y \neq d_p$. This completes the proof. □

Further, we observe that a function change similar to (5.2), i.e.

$$\begin{cases} \tilde{v}_k(x) = \tilde{u}_k(x) \cos \dfrac{\alpha^\sigma_k x}{d_k}, & k = \overline{1, n} \\[2mm] \tilde{v}_k(x) = \tilde{u}_k(x), & k = \overline{n+1, n+m} \end{cases} \tag{5.18}$$

converts the original steady state problem $(SE)+(SBC)$ into the abstract equation $A\tilde{u} = 0$ on the space $C(Y; \mathbf{R}^{n+m})$. This equation combined with (5.6) gives

$$\begin{cases} \dfrac{d}{dt}(u - \tilde{u}) = Au - A\tilde{u} \\[2mm] (u - \tilde{u})(0) = u_0 - \tilde{u} \ . \end{cases} \tag{5.19}$$

Now, Lemma 1.15 and Lemma 5.1 give

$$\frac{d^+}{dt} \|u(t, \cdot) - \tilde{u}(\cdot)\| \leq \omega^\sigma_\epsilon \|u(t, \cdot) - \tilde{u}(\cdot)\| \tag{5.20}$$

on $[0, \infty[$, where the $C(Y; \mathbf{R}^{n+m})$ norm was considered. Solving this differential inequality we obtain

$$\|u(t, \cdot) - \tilde{u}(\cdot)\| \leq \|u(0, \cdot) - \tilde{u}(\cdot)\| e^{\omega_\epsilon^\sigma \cdot t}$$

for all $t > 0$.

From here, by using (5.2) and (5.18) to return to the original variables v and $\tilde{v}$, and also taking into account the obvious inequalities

$$\min_{1 \leq i \leq n} \cos \gamma_i^\sigma \leq \cos \frac{\gamma_i^\sigma - \epsilon}{d_i} x_i \leq 1, \quad \text{for all } x_i \in [0, d_i],$$

we obtain

$$D(t) \leq e^{\omega_\epsilon^\sigma \cdot t} / \min_{1 \leq i \leq n} \cos \gamma_i^\sigma .$$

From here, if we take $\epsilon \to 0$ and denote

$$\omega_0^\sigma = \max \left\{ \max_{1 \leq j \leq n} \left[-\frac{(\gamma_j^\sigma)^2}{d_j^2 r_j c_j} - \frac{g_j}{c_j} \right]; \max_{n+1 \leq j \leq n+m} \left(\frac{-G_{n+j,n+j} + S_{n+j}}{s_{j-n}} \right) \right\} \quad (5.21)$$

we get

$$D(t) \leq \overline{D}_\sigma(t) = \frac{e^{\omega_0^\sigma \cdot t}}{\min_{1 \leq i \leq n} \cos \gamma_i^\sigma} \quad (5.22)$$

for all σ and $t \geq 0$.

Since from (5.20) we see that D is strictly decreasing ($\omega_\epsilon^\sigma < 0$), we find $T_\lambda = D^{-1}(\lambda)$. Also if we define

$$\overline{T}_\lambda(\sigma) = (\ln \lambda \min_{1 \leq i \leq n} \cos \gamma_i^\sigma) / \omega_0^\sigma = (\overline{D}_\sigma)^{-1}(\lambda),$$

the monotony of D^{-1} and $(\overline{D}_\sigma)^{-1}$ gives $T_\lambda \leq \overline{T}_\lambda(\sigma)$ for all σ. If in addition (5.1) is used, the above facts can be summarized as follows:

Theorem 5.1. *Let us consider the mixed type network from Figure 4.5 with a resistive multiport of $(G, B, 2n + m)$-type where G has the DRD property. Then,*

  i)   the delay $D : [0, \infty[\to \mathbf{R}$ is a strictly decreasing function,

 ii)   the direct current solution of our problem is globally exponential asymptotically stable, and

iii)   the upper bound of the λ-delay time is

$$\overline{T}_\lambda = \min_\sigma \frac{\ln \lambda \min_{1 \leq i \leq n} \cos \gamma_i^\sigma}{\omega_0^\sigma} , \quad (5.23)$$

where w_0^σ and $\cos \gamma_i^\sigma$ are given by (5.21) and (5.7) respectively.

The second statement above assures that, regardless of initial conditions, all dynamic solutions in the classical sense tend in $C(Y; \mathbf{R}^{n+m})$ to the same constant value, after the simultaneous connection of constant sources. In fact, in the same way we can obtain the stability of any solution (Marinov, Neittaanmäki [1988]) and its boundedness as well.

The third result above is an upper bound for λ-delay time. The formula (5.23) implies 2^n times application of relations (5.21) and (5.7), which involve all parameters of the circuit: r_i, c_i, g_i, d_i, s_i and G_{ij}. The simplicity of calculus makes this formula proper for fast simulators, used in digital network design. Of course, it is necessary that the upper bound is tight enough. This fact will be verified by the numerical computation of T_λ in Chapter 6 where several examples will be given.

5.3. A nonlinear mixed-type circuit

As we have seen above, the essential tool for delay time evaluation in linear mixed-type networks was the dissipativity of the operator governing the dynamic evolution. But the linearity of this operator plays no part in mathematical reasonings. From here derives the idea to extend the above approach to a nonlinear case.

Let us now consider the general network from Figure 5.1, where a $(G, B, 2p+2n)$-type resistive multiport connects p bipolar transistors $T_1 - T_p$ and n distributed parameter elements ("rcg-lines"), $L_1 - L_n$.

For the transistors we shall consider the nonlinear Gummel-model (presented in Figure 6.3) and reproduced in Figure 5.2 to specify some different notations:

According to physical reality, the transistor model contains the functions f_{2k-1}, $f_{2k} : \mathbf{R} \to \mathbf{R}$ with strictly positive derivatives and six strictly positive parameters $\alpha_{2k-1}, \alpha_{2k}, s_{2k-1}, s_{2k}, \tau_{3k-1}, \tau_{2k}$. From Figure 5.2 we easily derive

$$\begin{cases} u_{2k-1} = z_{2k-1} + \dfrac{\tau_{2k-1}}{s_{2k-1}} f_{2k-1}(z_{2k-1}) \equiv h_{2k-1}(z_{2k-1}) \\ u_{2k} = z_{2k} + \dfrac{\tau_{2k}}{s_{2k}} f_{2k}(z_{2k}) \equiv h_{2k}(z_{2k}) \end{cases} \tag{5.24}$$

$$\begin{bmatrix} i_{2k-1} \\ i_{2k} \end{bmatrix} = \begin{bmatrix} 1 & -\alpha_{2k} \\ -\alpha_{2k-1} & 1 \end{bmatrix} \begin{bmatrix} f_{2k-1}(z_{2k-1}) \\ f_{2k}(z_{2k}) \end{bmatrix} + \begin{bmatrix} s_{2k-1} \dfrac{du_{2k-1}}{dt} \\ s_{2k} \dfrac{du_{2k}}{dt} \end{bmatrix} \tag{5.25}$$

both valid for $k = \overline{1, p}$.

On the other hand, if we denote by $u_{2p+k}(t, x)$ and $i_{2p+k}(t, x)$ the voltage, and the current respectively, at the moment t and at the point $x \in]0, d_k[$ of the line L_k,

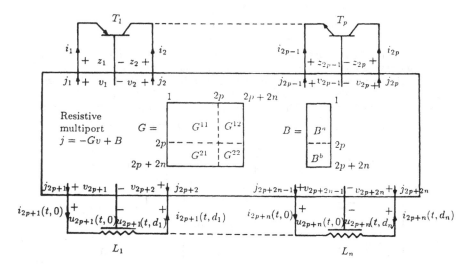

Figure 5.1 The network under study

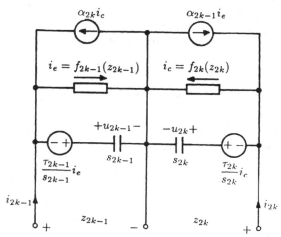

Figure 5.2 Large signal model of the kth transistor

the Telegraph Equations give:

$$(E) \qquad \begin{cases} \dfrac{\partial u_{2p+k}}{\partial t} = \dfrac{1}{r_k c_k} \dfrac{\partial^2 u_{2p+k}}{\partial x^2} - \dfrac{g_k}{c_k} u_{2p+k} \\[2mm] t \geq 0, \ x \in]0, d_k[, \ k = \overline{1,n} \ . \end{cases}$$

If we consider all sources having constant values (that is $B(t) = B$) and denote the terminal currents and voltages by $j = (j_1, \ldots, j_{2p+2n})^{tr}$ and $v = (v_1, \ldots, v_{2p+2n})^{tr}$ respectively, then the multiport imposes the constraint $j = -Gv + B$. Also, Figure 5.1 furnishes simple relations between the terminals variables i_k and j_k, z_k and v_k, u_k and v_k, respectively. Then, taking into account (5.24), (5.25) and noting that $-\frac{1}{r_k}\frac{\partial u_{2p+k}}{\partial x} = i_{2p+k}$ we obtain the following system of nonlinear boundary conditions:

$$(BC) \quad \begin{bmatrix} s_1 \dfrac{du_1}{dt}(t) \\ \vdots \\ s_{2p} \dfrac{du_{2p}}{dt} \\ -\dfrac{1}{r_1}\dfrac{\partial u_{2p+1}}{\partial x}(t,0) \\ +\dfrac{1}{r_1}\dfrac{\partial u_{2p+1}}{\partial x}(t,d_1) \\ \vdots \\ -\dfrac{1}{r_n}\dfrac{\partial u_{2p+n}}{\partial x}(t,0) \\ +\dfrac{1}{r_n}\dfrac{\partial u_{2p+n}}{\partial x}(t,d_n) \end{bmatrix} = -G \begin{bmatrix} h_1^{-1}(u_1(t)) \\ \vdots \\ h_{2p}^{-1}(u_{2p}(t)) \\ u_{2p+1}(t,0) \\ u_{2p+1}(t,d_1) \\ \vdots \\ u_{2p+n}(t,0) \\ u_{2p+n}(t,d_n) \end{bmatrix} - \begin{bmatrix} T\begin{bmatrix} f_1(h_1^{-1}(u_1(t))) \\ \vdots \\ f_{2p}(h_{2p}^{-1}(u_{2p}(t))) \end{bmatrix} \\ 0 \\ 0 \\ \vdots \\ 0 \end{bmatrix} + B,$$

$t > 0$, where $T = \bigoplus_{k=1}^{p} \begin{bmatrix} 1 & -\alpha_{2k} \\ -\alpha_{2k-1} & 1 \end{bmatrix}$.

We add the initial conditions $u_0 = (u_{1,0}, \ldots, u_{2p+n,0})^{tr}$

$$(IC) \qquad \begin{cases} u_k(0) = u_{k,0}, & k = \overline{1,2p} \\ u_{2p+k}(0,x) = u_{2p+k,0}(x), & x \in]0, d_k[, \ k = \overline{1,n} \ . \end{cases}$$

Some additional notations are necessary:

$$L_i = \max\left(-\frac{s_i}{\tau_i} \ ; \ -G_{ii}\right) + \sum_{\substack{k=1 \\ k \neq i}}^{2p} \max\left(\frac{|t_{ik}|s_k}{\tau_k} \ ; \ |G_{ik}|\right) + \sum_{k=2p+1}^{2p+n} |G_{ik}| \ , \qquad (5.26)$$

for $i = \overline{1,2p}$, where t_{ik} are elements of T. Our result will be derived supposing that the parameters satisfy:

A1 $\qquad \begin{cases} \text{For each } i = \overline{1,2p}, \ L_i < 0 \\ \text{For each } i = \overline{2p+1, 2p+n}, \ -G_{ii} + S_i < 0 \ . \end{cases}$

If we observe that $-G_{ii} + S_i \leq L_i$ for $i = \overline{1,2p}$, then we deduce that **A1** is (a little) more restrictive than the DRD property for the matrix G.

If **A1** is valid there exists a unique $\gamma_i \in]0, \pi/2[$ such that

$$\cos \gamma_i = \frac{S_{2p+2i} + \sqrt{S_{2p+2i}^2 + \dfrac{4}{r_i d_i}\left(G_{2p+2i,2p+2i} + \dfrac{1}{r_i d_i}\right)}}{2\left(G_{2p+2i,2p+2i} + \dfrac{1}{r_i d_i}\right)}, \quad i = \overline{1,n}. \tag{5.27}$$

Also, for every $\epsilon \geq 0$ with $\gamma_i - \epsilon \in]0, \pi/2[$, we denote

$$\omega_\epsilon = \max\left\{ \max_{1 \leq i \leq 2p} \frac{L_i}{S_i} \ ; \ \max_{1 \leq i \leq n}\left[-\frac{(\gamma_i - \epsilon)^2}{d_i^2 r_i c_i} - \frac{g_i}{c_i}\right]\right\}. \tag{5.28}$$

As in the preceding section, we shall consider all voltages as time and space variables $u_i : [0, \infty[\times [0, d_{i-2p}] \mapsto \mathbf{R}$ for $i = \overline{1, 2p+n}$ where $d_{i-2p} = 0$ for $i = \overline{1,2p}$. Let $Y = \prod_{i=1}^{2p+n} [0, d_{i-2p}]$. We suppose that for $u_0 \in C(Y; \mathbf{R}^{2p+n})$ with $u_{k,0} \in C^2(]0, d_k[, \mathbf{R})$ there exists a solution in the classical sense of $(E) + (BC) + (IC)$, $u \in C(Y; \mathbf{R}^{2p+n})$ with $u_k(t) \in C^2(]0, d_k[, \mathbf{R})$ for $t \geq 0$. Also, we suppose the existence of a steady state solution $\tilde{u} = (\tilde{u}_1, \ldots, \tilde{u}_{2p+n})^{tr} \in C(Y; \mathbf{R}^{2p+n})$ where $\tilde{u}_k \in C^2(]0, d_k[, \mathbf{R})$.

If we define, as in Section 5.1, "the delay" as $D : [0, \infty[\to \mathbf{R}$, $D(t) = \|u(t) - \tilde{u}\| / \|u_0 - \tilde{u}\|$ where the norm of space $C(Y; \mathbf{R}^{2p+n})$ was used, then we can prove the following result:

Theorem 5.2. *Suppose that for the problem* $(E) + (BC) + (IC)$, $u_{i,0} \neq \tilde{u}_i$ *for every* $i = \overline{1, 2p+n}$, *and that* **A1** *is valid. Then* $\lim_{t \to \infty} u_i(t, x) = \tilde{u}_i(x)$ *uniformly in* x *for* $i = \overline{1, 2p+n}$. *Moreover, the delay is a strictly decreasing function and*

$$D(t) \leq \overline{D}(t) = e^{\omega_0 t} / \min_{1 \leq i \leq n} \cos \gamma_i, \tag{5.29}$$

where w_0 *and* $\cos \gamma_i$ *are given by (5.28) and (5.27) respectively.*

Proof. The method of proof will be the same as above: we formulate the problem as a Cauchy one in $C(Y; \mathbf{R}^{2p+n})$ and the assymptotic stability will be implied by a dissipative property.

First of all, a change of functions appearing in $(E) + (BC) + (IC)$:

$$\begin{cases} u_k(t, x) = w_k(t, x), & k = \overline{1, 2p} \\[2mm] u_k(t, x) = w_k(t, x)\dfrac{\cos \beta_{k-2p} x}{d_{k-2p}}, & k = \overline{2p+1, 2p+n}, \end{cases} \tag{5.30}$$

where β_{k-2p} will be choosen later for each k. We consider an operator A : $\mathcal{D}(A) \subset C(Y; \mathbf{R}^{2p+n}) \to C(Y; \mathbf{R}^{2p+n})$, where $\mathcal{D}(A)$ comprises the functions $w \in C(Y; \mathbf{R}^{2p+n})$ with $w_i \in C^2(]0, d_{i-2p}[; \mathbf{R})$ and satisfying

$$
\begin{bmatrix} -\dfrac{1}{r_1}\dfrac{dw_{2p+1}}{dx}(0) \\[2mm] \dfrac{1}{r_1}\dfrac{dw_{2p+1}}{dx}(d_1)\cos\beta_1 \\[2mm] \vdots \\[2mm] -\dfrac{1}{r_1}\dfrac{dw_{2p+n}}{dx}(0) \\[2mm] \dfrac{1}{r_1}\dfrac{dw_{2p+n}}{dx}(d_n)\cos\beta_n \end{bmatrix} = -G^{21}\begin{bmatrix} h_1^{-1}(w_1(0)) \\[2mm] \vdots \\[2mm] h_{2p}^{-1}(w_{2p}(0)) \end{bmatrix} - G^{22}\begin{bmatrix} w_{2p+1}(0) \\[2mm] w_{2p+1}(d_1)\cos\beta_1 \\[2mm] \vdots \\[2mm] w_{2p+n}(0) \\[2mm] w_{2p+n}(d_n)\cos\beta_n \end{bmatrix}
$$

$$
+ \begin{bmatrix} 0 \\[2mm] \dfrac{\beta_1 w_{2p+1}(d_1)}{r_1 d_1}\sin\beta_1 \\[2mm] \vdots \\[2mm] 0 \\[2mm] \dfrac{\beta_n w_{2p+n}(d_n)}{r_n d_n}\sin\beta_n \end{bmatrix} + B^b.
$$

The operator A is defined as follows:

- the first $2p$ components are:

$$
\begin{bmatrix} (Aw)_1(x) \\[2mm] \vdots \\[2mm] (Aw)_{2p}(x) \end{bmatrix} = \operatorname{diag}(s_1^{-1}, \ldots, s_{2p}^{-1})\left\{ -G^{11}\begin{bmatrix} h_1^{-1}(w_1(x_1)) \\[2mm] \vdots \\[2mm] h_{2p}^{-1}(w_{2p}(x_{2p})) \end{bmatrix} - \right.
$$

$$
\left. -G^{12}\begin{bmatrix} w_{2p+1}(0) \\[1mm] w_{2p+1}(d_1)\cos\beta_1 \\[1mm] \vdots \\[1mm] w_{2p+n}(0) \\[1mm] w_{2p+n}(d_n)\cos\beta_n \end{bmatrix} - T\begin{bmatrix} f_1(h_1^{-1}(w_1(x_1))) \\[2mm] \vdots \\[2mm] f_{2p}(h_{2p}^{-1}(w_{2p}(x_{2p}))) \end{bmatrix} + B^a \right\}
$$

- the last n components are:

$$
(Aw)_{2p+k}(x) = \frac{1}{r_k c_k}\frac{d^2 w_{2p+k}}{dx^2} - \left(\frac{2\beta_k}{r_k c_k d_k}\tan\frac{\beta_k x_{2p+k}}{d_k}\right)\frac{dw_{2p+k}}{dx} -
$$

$$
- \left(\frac{\beta_k^2}{d_k^2 r_k c_k} + \frac{g_k}{c_k}\right)w_{2p+k} \quad \text{for } k = \overline{1, n} .
$$

It is straightforward to show that $(E) + (BC) + (IC)$ together with (5.30) give the following differential equation in $C(Y; \mathbf{R}^{2p+n})$:

$$\begin{cases} \dfrac{dw(t,\cdot)}{dt} = Aw(t,\cdot) \\ w_k(0,\cdot) = \begin{cases} u_{k,0} & \text{for } k = \overline{1,2p} \\ u_{k,0}(\cdot)/\cos\dfrac{\beta_{k-2p}\cdot(\cdot)}{d_k - 2p} & \text{for } k = \overline{2p+1, 2p+n} \ . \end{cases} \end{cases} \tag{5.31}$$

Let us take $w, \overline{w} \in \mathcal{D}(A)$ with $\psi = w - \overline{w}$ and

$$M(\psi) = \Big\{ (q;y) \mid q = \overline{1, 2p+n}, \ y \in [0, d_{q-2p}],$$

$$\max_{2p+1 \leqslant i \leqslant 2p+n} \max_{x \in [0, d_{i-2p}]} |\psi_i(x)| = |\psi_q(y)| = \|\psi\| \Big\} \ .$$

For $(q;y) \in M(\psi)$ and $q = \overline{1, 2p}$ we obtain

$$[Aw - A\overline{w}]_q(y)\,\mathrm{sgn}\,\psi_q(y) \leq$$

$$-\frac{1}{s_q}\Big[\sum_{k=1}^{2p} \frac{t_{qk}f'_k + G_{qk}}{s_k + \tau_k f'_k} s_k \psi_k(y)\,\mathrm{sgn}\,\psi_q(y) + \sum_{k=2p+1}^{2p+2n} |G_{qk}||\psi_q(y)|\Big] \ ,$$

where we have applied the mean value theorem and denoted by f'_k the (positive) derivative of function f_k in the intermediate point from $]h_k^{-1}(w_k), h_k^{-1}(\overline{w}_k)[$. So, for $q = \overline{1, 2p}$ we have

$$[Aw - A\overline{w}]_q(y)\,\mathrm{sgn}\,\psi_q(y) \leq \frac{L_q}{s_k}|\psi_q(y)| \ . \tag{5.32}$$

For $(q;y) \in M(\psi)$ with $q = \overline{2p+1, 2p+n}$, we choose $\beta_j = \gamma_j - \epsilon$ where γ_j is given by (5.27) and $\epsilon > 0$ is such that $\gamma_j - \epsilon \in]0, \pi/2[$. Reasoning as in the proof of Lemma 5.1 (we omit the details) we obtain

$$[Aw - A\overline{w}]_q(y)\,\mathrm{sgn}\,\psi_q(y) \leq -\left(\frac{\beta_{q-2p}^2}{r_{q-2p}c_{q-2p}d_{q-2p}^2} + \frac{g_{q-2p}}{c_{q-2p}}\right)|\psi_q(y)| \ . \tag{5.33}$$

Now, (5.32) and (5.33) give the total dissipativity of A (see Lemma 1.6) for all above chosen ϵ:

$$\langle Aw - A\overline{w}, w - \overline{w}\rangle_+ \leq \omega_\epsilon \|w - \overline{w}\|.$$

If we remark that **A1** implies $\omega_\epsilon < 0$, from here we derive the result. $\qquad \square$

If we adopt for the λ-delay time the same definition as in preceding section, from (5.29) we derive

$$T_\lambda \leq \overline{T}_\lambda = (\ln\lambda \min_{1 \leqslant i \leqslant n} \cos\gamma_i)/\omega_0 \ , \tag{5.34}$$

where $\cos\gamma_i$ and ω_0 are easily computable (see (5.27) and (5.28)).

5.4. Comments

The stability of the circuits with distributed structures has been studied by many authors using different methods. Prada and Bickart [1971] use a Lyapunov theory for a functional-differential equation of retarded type that describes a large class of such circuits. A small signal stability criterion (in terms of the roots of characteristic equations) is derived in Brayton [1968] while input-output stability results are given in Desoer [1977] for circuits with parasitic elements, by applying the theory of singular perturbations.

The extensive bibliographies of Ghausi and Kelly [1968] and Kumar [1980] summarize the work on distributed rc-circuits up to 1980. We have to remark that for a nonuniform open circuited rc-line, Protonotarios and Wing [1967] show the step response to be monotonously increasing and Singhal and Vlach [1972] obtain bounds of this response. As regards the delay time, a pioneering work in this field is Elmore [1948] who called the first moment of the impulse response the delay.

The explosive increase in the work on transmission lines in the past decade has been mainly motivated by the preoccuppation with the delay time in MOS interconnections (see Section 4.0) and for the performances of microwave transmissions. Some authors have worked on the transient analysis of a single transmission line giving exact analytical time domain expressions for voltage and current at any point on the line: Cases and Quinn [1980], Preis and Shlager [1988], Zurada and Liu [1987].

Many other authors (Gao et al. [1990], Passlack et al. [1990], Araki and Naito [1985]) tried to implement a transmission line model in a general purpose CAD circuit simulator such as SPICE. But, for delay time prediction at the design stage, a much faster simulator is needed (see Section 6.0). To this goal, one of the most accepted methods is to use a very simple RC lumped model of the whole network as a basis to infer easily computable bounds of the delay time. The first result in this direction is due to Rubinstein, Penfield and Horowitz [1983] on RC "tree" networks. Extensions of these bounds to nonlinear RC networks and to RC mesh networks were given by Wyatt [1985 a,b]. An interesting extension of Elmore's delay to RC networks was found by Chan [1986 a,b]. Bounds which can be improved iteratively were developed by Zukowski [1986 a,b] while the delay time sensitivity is treated in Jain et all. [1987]. Also, RC mesh type circuits have been studied by Lin and Mead [1984], Chan and Schlag [1989], Harbour and Drake [1989]. Relating to all these papers, we observe that the accuracy of approximation of interconnections by RC ladder networks is not clear, being studied only for one line: Sakurai [1983]. For instance, for a open circuited rc-line of length d, the rise time is rcd^2, while for a RC cell with $R = rd$ and $C = cd$ the rise time is 2.3 rcd^2. That is why our bound of the delay time in a network with exactly modelled rcg-lines is probably welcome. Of course, a lower bound of the same type is desirable. We have done it recently but working in a completely different way, Marinov and Neittaanmäki [1991 a,b].

On the other hand, we have to observe that the delay time notion in all the above papers is related to a given input-output path, while our delay time is a global one, describing the speed of signal propagation after switching of a part or all sources. Of course, one of our bound shortcomings is the a-priori necessity to describe the resistive part of the network as a $(G, B, 2n + m)$ multiport.

Finally, let us mention that the results of this chapter are obtained in Marinov [1987], Marinov and Neittaanmäki [1986, 1988, 1989, 1990 b]. Another nonlinear case can be found in Marinov and Neittaanmäki [1990 a].

Chapter VI

Numerical approximation of mixed models for digital integrated circuits

6.0. Introduction

To analyse an electrical network many CAD (Computer Aided Design) circuit simulators are available today. The most well-known is probably SPICE –Nagel [1975]. Although this type of simulator is able to precisely compute the transient performances (as delay time), the usage of complete models of devices implies an extremely high time consumption. So, the circuit simulators are unappropriate for the initial stage of VLSI design where a high speed timing analyser ("timing simulator") is required. To this goal, alternative approaches using either simpler device models or simpler numerical algorithms or easily computable formulae for delay time approximation, have been developed in the past decade to improve the simulation efficiency: Terman [1985], Ousterhout [1985], White and Sangiovanni-Vincentelli [1986], Kim [1986], Putatunda [1984], Tsao and Chen [1986], Jouppi [1987], Lin and Mead [1986], Pillage and Rohrer [1990], Chan and Karplus [1990]. Thus, timing analysers (e.g. Ousterhout [1985], Jouppi [1987]) are often able to predict the interconnect delay with a simplified model (typically an RC tree) to within 10 percent of a SPICE prediction, by using a much shorter simulation time (Pillage and Rohrer [1990]).

Below, we have built a simulator for an integrated chip by using our mixed-type model. The fact that the interconnection lines are modelled by Telegraph Equations makes our program proper for wiring delay computation and more precise than other timing simulators where the lines are lumped modelled. The space discretization is based on a variational formulation of the problem and on the use of the finite element method. Thus, it is easy and convenient to handle the "crossed" boundary conditions. The semidiscrete model (called FEM-model below) leads to an initial value problem for a system of differential equations. Typically this system is stiff and we shall apply NAG-subroutine to solve it numerically. Several examples are given in Section 6.7. On this occasion we shall numerically verify the bound infered in the previous chapter for the delay time.

Previous partial treatment of the numerical approach to our problem was given in Marinov, Neittaanmäki and Hara [1987], Marinov and Neittaanmäki [1986, 1988, 1989] and Neittaanmäki, Hara and Marinov [1988].

6.1. The mathematical model

Let us consider again the general mixed-type circuit of Figure 4.5 which has n lines and m capacitors connected to a resistive multiport. The mathematical model comprises of

– a system of parabolic equations for the voltage $v = (v_1, ..., v_n)$ on lines

$$\frac{\partial v_k}{\partial t} = \frac{1}{r_k c_k} \frac{\partial^2 v_k}{\partial x^2} - \frac{g_k}{c_k} v_k, \quad t \geq 0, \ x \in]0, d_k[, \ k = \overline{1, n}, \tag{6.1}$$

– a system of "crossed" boundary conditions

$$\begin{bmatrix} -\dfrac{1}{r_1} \dfrac{\partial v_1(t, 0)}{\partial x} \\ \dfrac{1}{r_1} \dfrac{\partial v_1(t, d_1)}{\partial x} \\ \vdots \\ -\dfrac{1}{r_n} \dfrac{\partial v_n(t, 0)}{\partial x} \\ \dfrac{1}{r_n} \dfrac{\partial v_n(t, d_n)}{\partial x} \\ i_{2n+1}(t) \\ \vdots \\ i_{2n+m}(t) \end{bmatrix} = -G \begin{bmatrix} v_1(t, 0) \\ v_1(t, d_1) \\ \vdots \\ v_n(t, 0) \\ v_n(t, d_n) \\ v_{n+1}(t) \\ \vdots \\ v_{n+m}(t) \end{bmatrix} + B(t), \quad t \geq 0, \tag{6.2}$$

– a set of given initial conditions

$$\begin{cases} v_k(0, x) = v_{k0}(x), & k = 1, \ldots, n \ , \\ v_{n+k}(0) = v_{n+k,0}, & k = 1, \ldots, m \ . \end{cases} \tag{6.3}$$

In general, the current, the voltage and the electric charge of the kth capacitor are related by

$$\begin{cases} v_{n+k}(t) = f_k^{CAP}(q_k(t)) \\ i_{2n+k}(t) = \dfrac{dq_k(t)}{dt} \ . \end{cases} \tag{6.4}$$

Taking into account that our theory in preceeding chapters was given for the linear case, we shall describe below only this case, namely

$$f_k^{CAP}(q_k) = \frac{q_k}{s_k} \tag{6.5}$$

where s_k is capacitance. (Let us mention that our simulator is more general, and also works for the nonlinear case.) It follows

$$i_{2n+k}(t) = s_k \frac{dv_{n+k}}{dt}(t) \tag{6.6}$$

and from (6.2) and (6.6) we can derive to line $k \in \overline{1,n}$

$$\begin{cases} -\dfrac{1}{r_k} \dfrac{\partial v_k(t,0)}{\partial x} = -G_{2k-1}u(t) + B_{2k-1}(t) \\ \dfrac{1}{r_k} \dfrac{\partial v_k(t,d_k)}{\partial x} = -G_{2k}u(t) + B_{2k}(t) \end{cases} \tag{6.7}$$

and to linear capacitor $k \in \overline{1,n}$

$$s_k \frac{dv_{n+k}(t)}{dt} = -G_{2n+k}u(t) + B_{2n+k}(t). \tag{6.8}$$

In (6.7) and (6.8) we have used the following notations: G_j denotes the jth row of the matrix G of boundary conditions, $B_j(t)$ denotes the jth component of the vector $B(t)$ and the voltages at the end points of lines and in capacitors are denoted by

$$u(t) = (u_1(t), \ldots, u_{2n+m}(t))^{tr}$$
$$= \Big(\underbrace{v_1(t,0), v_1(t,d_1)}_{\text{first line}}, \ldots, \underbrace{v_n(t,0), v_n(t,d_n)}_{\text{last line}},$$
$$\underbrace{v_{n+1}(t)}_{\text{first capacitor}}, \ldots, \underbrace{v_{n+m}(t)}_{\text{last capacitor}} \Big)^{tr} . \tag{6.9}$$

Also it is convenient to partition the matrix G as follows:

$$G = (G_{ij})_{i,j=1}^{2n+m}$$
$$= \begin{bmatrix} G_{11}^{(2n),(2n)} & G_{1,(2n+1)}^{(2n),(2n+m)} \\ G_{(2n+1),1}^{(2n+m),(2n)} & G_{(2n+1),(2n+1)}^{(2n+m),(2n+m)} \end{bmatrix} \tag{6.10}$$

with blocks

$$G_{ij}^{kl} = \begin{bmatrix} G_{ij} & \ldots & G_{il} \\ \vdots & \ddots & \vdots \\ G_{kj} & \ldots & G_{kl} \end{bmatrix} . \tag{6.11}$$

6.2. Construction of the system of FEM-equations

The system of the differential equations in time is obtained from the partial differential equations of lines and the differential equations of the boundary conditions by using the following principles:

- FEM-model is obtained by discretizing the lines in space by using the finite element method. If line k $(k = 1, \ldots, n)$ has N_k discretization intervals ($N_k + 1$ discretization points) we get $\sum_{k=1}^{n}(N_k + 1)$ differential equations
- The differential equations (6.8) describing capacitors are added to the FEM-model for lines. Consequently, the problem (6.1)–(6.3) leads to an initial value problem with $\sum_{k=1}^{n}(N_k + 1) + m$ differential equations.

6.2.1. Space discretization of rcg-lines

Let us take the ith $(i = 0, \ldots, N_k)$ basis function of line k $(k = 1, \ldots, n)$, $\phi_i : [0, d_k] \mapsto \mathbf{R}$ defined by (see Figure 6.1)

$$\phi_i(x) = \begin{cases} \dfrac{x}{\Delta h_k} - (i - 1), & \text{if} \quad (i-1)\Delta h_k \leq x \leq i\Delta h_k \\[2mm] -\dfrac{x}{\Delta h_k} + (i + 1), & \text{if} \quad i\Delta h_k \leq x \leq (i+1)\Delta h_k \\[2mm] 0, & \text{elsewhere} \end{cases} \tag{6.12}$$

which is used in a space discretization of the partial differential equations of lines.

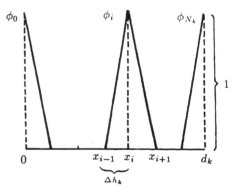

Figure 6.1

Here we have denoted by

$$\Delta h_k = \frac{d_k}{N_k} \tag{6.13}$$

the length of the discretization interval. So the values of basis functions at the discretization points are as follows

$$\phi_i(x_j) = \phi_i(j\Delta h_k) = \begin{cases} 1, & \text{if } i = j \\ 0, & \text{if } i \neq j \end{cases}.$$

If we multiply the partial differential equation (6.1) with the basis function $\phi_i(x)$ and integrate over line k we get

$$\int_0^{d_k} \frac{\partial v_k(t,x)}{\partial t} \phi_i(x) dx = \frac{1}{r_k c_k} \int_0^{d_k} \frac{\partial^2 v_k(t,x)}{\partial x^2} \phi_i(x) dx \\ - \frac{g_k}{c_k} \int_0^{d_k} v_k(t,x) \phi_i(x) dx \tag{6.14}$$

which, by applying Green's formula, gives

$$\frac{\partial}{\partial t} \int_0^{d_k} v_k(t,x) \phi_i(x) dx = \frac{1}{r_k c_k} \Bigg|_0^{d_k} \frac{\partial v_k(t,x)}{\partial x} \phi_i(x) \\ - \frac{1}{r_k c_k} \int_0^{d_k} \frac{\partial v_k(t,x)}{\partial x} \frac{\partial \phi_i(x)}{\partial x} dx \tag{6.15} \\ - \frac{g_k}{c_k} \int_0^{d_k} v_k(t,x) \phi_i(x) dx.$$

Taking into account the boundary conditions of line (6.7) we obtain

$$\frac{1}{r_k c_k} \Bigg|_0^{d_k} \frac{\partial v_k(t,x)}{\partial x} \phi_i(x)$$

$$= \frac{1}{c_k} \left(\frac{1}{r_k} \frac{\partial v_k(t,d_k)}{\partial x} \phi_i(d_k) - \frac{1}{r_k} \frac{\partial v_k(t,0)}{\partial x} \phi_i(0) \right)$$

$$= \frac{1}{c_k} \left(-G_{2k} u(t) + B_{2k}(t) \right) \phi_i(d_k)$$

$$+ \frac{1}{c_k} \left(-G_{2k-1} u(t) + B_{2k-1}(t) \right) \phi_i(0)$$

$$\equiv f_{ki} \equiv \begin{cases} -\dfrac{1}{c_k} G_{2k-1} u(t) + \dfrac{1}{c_k} B_{2k-1}(t), & \text{if } i = 0 \\ 0, & \text{if } i \neq 0 \text{ and } i \neq N_k, \ i = 1, ..., N_{k-1} \\ -\dfrac{1}{c_k} G_{2k} u(t) + \dfrac{1}{c_k} B_{2k}(t), & \text{if } i = N_k \end{cases}.$$

$$\tag{6.16}$$

As we see, ϕ_i joins the boundary conditions to the differential equations only at the first (start of line) and the last (end of line) discretization points of each line. We approximate the voltage $v_k(t, x)$ of the line k by

$$v_{h,k}(t, x) = \sum_{j=0}^{N_k} v_{h,k}^j(t)\phi_j(x) \tag{6.17}$$

where $v_{h,k}^j(t)$ is the voltage at the jth discretization point x_j and at time t. This approximation is of order $\mathcal{O}(h_k^2)$ (see Křížek-Neittaanmäki [1990], for example).

If we replace in (6.15) v by v_h we get

$$\sum_{j=0}^{N_k} \frac{\partial v_{h,k}^j(t)}{\partial t} \underbrace{\int_0^{d_k} \phi_i(x)\phi_j(x)dx}_{\equiv a_{ij}^k} = \frac{1}{r_k c_k} \sum_{j=0}^{N_k} v_{h,k}^j(t) \underbrace{\int_0^{d_k} \frac{\partial \phi_i(x)}{\partial x} \frac{\partial \phi_j(x)}{\partial x} dx}_{\equiv b_{ij}^k}$$

$$- \frac{g_k}{c_k} \sum_{j=0}^{N_k} v_{h,k}^j(t) \underbrace{\int_0^{d_k} \phi_i(x)\phi_j(x)dx}_{\equiv a_{ij}^k} + f_{kj} \quad . \tag{6.18}$$

If we calculate the integrals a_{ij}^k and b_{ij}^k in equation (6.18) for all combinations of i and j we get for line k

$$\widehat{A}_k = (a_{ij}^k)_{i,j=0}^{N_k} = \frac{\Delta h_k}{6} \begin{bmatrix} 2 & 1 & & & \\ 1 & 4 & 1 & & \\ & \ddots & \ddots & \ddots & \\ & & 1 & 4 & 1 \\ & & & 1 & 2 \end{bmatrix} \tag{6.19}$$

and

$$\widehat{B}_k = (b_{ij}^k)_{i,j=0}^{N_k} = \frac{1}{\Delta h_k} \begin{bmatrix} 1 & -1 & & & \\ -1 & 2 & -1 & & \\ & \ddots & \ddots & \ddots & \\ & & -1 & 2 & -1 \\ & & & -1 & 1 \end{bmatrix} \quad . \tag{6.20}$$

6.2.2. FEM-equations of lines

The differential equations (6.18) of lines obtained by FEM discretization can be presented in the form

$$MV_L'(t) + (\overline{M} + K + \widetilde{G}_L)V_L(t) = \widetilde{B}_L(t) \tag{6.21_1}$$

with the initial condition

$$V_L(0) = V_{0L} , \tag{6.21$_2$}$$

where

$$M = \begin{bmatrix} \widehat{A}_1 & & \\ & \ddots & \\ & & \widehat{A}_n \end{bmatrix} , \tag{6.22}$$

$$\overline{M} = \begin{bmatrix} \dfrac{g_1}{c_1}\widehat{A}_1 & & \\ & \ddots & \\ & & \dfrac{g_n}{c_n}\widehat{A}_n \end{bmatrix} , \tag{6.23}$$

and

$$K = \begin{bmatrix} \dfrac{1}{r_1 c_1}\widehat{B}_1 & & \\ & \ddots & \\ & & \dfrac{1}{r_n c_n}\widehat{B}_n \end{bmatrix} . \tag{6.24}$$

The sizes of these square matrices are

$$N(n) = \sum_{k=1}^{n}(N_k + 1) . \tag{6.25}$$

The vector $V_L(t)$ contains the values of voltages $v_k(t, j\Delta h_k)$ ($j = 0, ... N_k$ and $k = 1, ..., n$) at all discretization points of lines at time t and its size is also $N(n)$ and $V_{0L} = (v_{01}, ... v_{0N(n)})$ is the pointwise initial condition corresponding to condition (6.3). The square matrix $\widetilde{G}_L$ (size $N(n) \times N(n)$) has the following form

$$\widetilde{G}_L = (\widetilde{G}_{ij})_{i,j=1}^{N(n)} = \begin{bmatrix} \widetilde{G}_{11}^{\text{LL}} & \dots & \widetilde{G}_{1n}^{\text{LL}} \\ \vdots & \ddots & \vdots \\ \widetilde{G}_{n1}^{\text{LL}} & \dots & \widetilde{G}_{nn}^{\text{LL}} \end{bmatrix} \tag{6.26}$$

where $\widetilde{G}_{ij}^{\text{LL}}$ (size $(N_i + 1) \times (N_j + 1)$) is a submatrix of the form

$$\widetilde{G}_{ij}^{\text{LL}} = \frac{1}{c_i} \begin{bmatrix} G_{(2i-1),(2j-1)} & 0 & \dots & 0 & G_{(2i-1),(2j)} \\ 0 & 0 & \dots & 0 & 0 \\ \vdots & \vdots & \ddots & \vdots & \vdots \\ 0 & 0 & \dots & 0 & 0 \\ G_{(2i),(2j-1)} & 0 & \dots & 0 & G_{(2i),(2j)} \end{bmatrix} . \tag{6.27}$$

The corner terms of $\widetilde{G}_{ij}^{\text{LL}}$ correspond to the start and the end points of the lines and they are elements of matrix G.

The source vector (size $N(n)$) can be written

$$\widetilde{B}_L(t) = \left(\frac{B_1}{c_1}, 0, \ldots, 0, \frac{B_2}{c_1}, \ldots, \right.$$

$$\underbrace{\frac{B_{2k-1}}{c_k}, 0, \ldots, 0, \frac{B_{2k}}{c_k}}_{\text{line } k}, \ldots,$$

$$\left. \frac{B_{2n-1}}{c_n}, 0, \ldots, 0, \frac{B_{2n}}{c_n} \right)^{tr} . \tag{6.28}$$

6.3. FEM-equations of the model

Let us join the capacitors equations (6.8) to the system of FEM equations of lines (6.21_1). We get

$$\begin{bmatrix} M & \\ & S^{\text{CAP}} \end{bmatrix} V'(t) + \left(\begin{bmatrix} \overline{M} + K & \\ & O^{\text{CAP}} \end{bmatrix} + \widetilde{G} \right) V(t) = \widetilde{B}(t) \tag{6.29_1}$$

with the corresponding initial condition

$$V(0) = V_0 \tag{6.29_2}$$

where S^{CAP} (size m) is a diagonal matrix with capacitances s_i of capacitors and O^{CAP} (size m) is a zero square matric. The square matrix $\widetilde{G}$ can be written as

$$\widetilde{G} = (\widetilde{G}_{ij})_{i,j=1}^{N(n)+m} = \begin{bmatrix} \widetilde{G}^{\text{LL}} & \widetilde{G}^{\text{LC}} \\ \widetilde{G}^{\text{CL}} & \widetilde{G}^{\text{CC}} \end{bmatrix}$$

$$= \begin{bmatrix} \widetilde{G}_{11}^{\text{LL}} & \cdots & \widetilde{G}_{1n}^{\text{LL}} & \widetilde{G}_{11}^{\text{LC}} & \cdots & \widetilde{G}_{1m}^{\text{LC}} \\ \vdots & \ddots & \vdots & \vdots & \ddots & \vdots \\ \widetilde{G}_{n1}^{\text{LL}} & \cdots & \widetilde{G}_{nn}^{\text{LL}} & \widetilde{G}_{n1}^{\text{LC}} & \cdots & \widetilde{G}_{nm}^{\text{LC}} \\ \widetilde{G}_{11}^{\text{CL}} & \cdots & \widetilde{G}_{1n}^{\text{CL}} & \widetilde{G}_{11}^{\text{CC}} & \cdots & \widetilde{G}_{1m}^{\text{CC}} \\ \vdots & \ddots & \vdots & \vdots & \ddots & \vdots \\ \widetilde{G}_{m1}^{\text{CL}} & \cdots & \widetilde{G}_{mn}^{\text{CL}} & \widetilde{G}_{m1}^{\text{CC}} & \cdots & \widetilde{G}_{mm}^{\text{CC}} \end{bmatrix} , \tag{6.30}$$

where $\widetilde{G}^{\text{LC}}$, $\widetilde{G}^{\text{CL}}$ and $\widetilde{G}^{\text{CC}}$ come from the boundary condition terms related to capacitors.

Let us first consider submatrices which come from the boundary conditions of the lines. Line connections are the same as for matrix $\widetilde{G}_{ij}^{\text{LL}}$ in (6.27). We can write lines connection with the capacitors (a vertical vector whose size is $N_i + 1$; $i = 1, \ldots, n$ and $j = 1, \ldots, m$)

$$\widetilde{G}_{ij}^{LC} = \frac{1}{c_i} \left(G_{(2i-1),(2n+j)}, 0, \ldots, 0, G_{(2i),(2n+j)} \right)^{tr} . \tag{6.31}$$

Also, the terms corresponding to the connections between capacitors and lines (a horizontal vector whose size is $N_j + 1$; $i = 1, \ldots, m$ and $j = 1, \ldots, n$) are grouped together in

$$\widetilde{G}_{ij}^{CL} = \left(G_{(2n+i),(2j-1)}, 0, \ldots, 0, G_{(2n+i),(2j)} \right) . \tag{6.32}$$

Finally, a block corresponds to capacitor–capacitor (one term; $i = 1, \ldots, m$ and $j = 1, \ldots, m$) connection is defined by

$$\widetilde{G}_{ij}^{CC} = G_{(2n+i),(2n+j)} . \tag{6.33}$$

We can write the source vector (size $N(n) + m$), see (6.28))

$$\widetilde{B}(t) = \Bigg(\underbrace{\frac{B_1}{c_1}, 0, \ldots, 0, \frac{B_2}{c_1}, \ldots, \frac{B_{2n-1}}{c_n}, 0, \ldots, 0, \frac{B_{2n}}{c_n},}_{\widetilde{B}_L(t) \text{ (lines)}}$$

$$\underbrace{B_{2n+1}, \ldots, B_{2n+m}}_{\text{capacitors}} \Bigg)^{tr} \tag{6.34}$$

and the voltage vector (size $N(n) + m$)

$$V(t) = \Bigg(\underbrace{V_1(t), \ldots, V_{N(n)}(t)}_{V_L(t) \text{ (lines)}}, \underbrace{V_{N(n)+1}(t), \ldots, V_{N(n)+m}(t)}_{\text{capacitors}} \Bigg)^{tr}$$

$$= \Big(v_1(t,0), \ldots, v_1(t, j_1 \Delta h_1), \ldots, v_1(t, d_1), \ldots, \tag{6.35}$$

$$v_n(t,0), \ldots, v_n(t, j_n \Delta h_n), \ldots, v_n(t, d_n),$$

$$v_{n+1}(t), \ldots, v_{n+m}(t) \Big)^{tr},$$

where $j_k = 0, \ldots, N_k$ ($k = 1, \ldots, n$).

6.4. Residual evaluations

When the FEM is applied to the model (6.1), (6.2), (6.3), we obtained the initial value problem as described above. There are several integrators available for solving the numerically obtained problem (in the simplest case we could apply implicit Euler or Crank-Nicholson methods with a fixed time step). We have applied the subroutine D02NGF of NAG Fortran Library. D02NGF is a general purpose routine for integrating the initial value problems of stiff systems of implicit ordinary differential equations

$$A(t, y)y' = g(t, y) \tag{6.36}$$

and it has been developed from SPRINT package (Software for problems in time, Brezins, Dew, Furzeland [1989]).

Because in the case of this chapter, $A(t, y)$ does not depend on t and y, one could use the integrators designed for solving stiff systems of explicitly defined ordinary differential equations

$$y' = A^{-1}g(t, y). \tag{6.37}$$

Consequently, other subroutine like D02NBF and D02NDF of NAG could as well be applied, instead of D02NGF. In Appendix I we have outlined in detail the basic steps on how to solve the obtained initial value problem by integrator D02NGF.

The user-supplied RESID routine of D02NGF defines the system of differential-algebraic equations to be solved. The integrator supplies approximate vectors for the solution y and its time derivative y'. The main purpose of the RESID routine is to compute the residual vector $\mathbf{r}$

$$\mathbf{r} = g(t, y) - A(t, y)y' . \tag{6.38}$$

In residual evaluations the system of FEM differential equations is given in our case in the form

$$R(t) = -\begin{bmatrix} M & \\ & S^{\text{CAP}} \end{bmatrix} V'(t)$$
$$- \left(\begin{bmatrix} \overline{M} + K & \\ & O^{\text{CAP}} \end{bmatrix} - \widetilde{G} \right) V(t) + \widetilde{B}(t) \tag{6.39}$$

with the initial condition $V(0) = V_0$. The vector $R(t)$ is the value of the residual at the discretization points and at time t.

Let us note by $N(k)$ the number of the discretization points from the first line to line k ($N(0) = 0$) and use the abbreviations (see (6.18), (6.19) and (6.22) – (6.24))

$$\widehat{a}_k = \frac{2}{\Delta h_k r_k c_k} + \frac{4\Delta h_k g_k}{6c_k}$$
$$\widehat{b}_k = -\frac{1}{\Delta h_k r_k c_k} + \frac{\Delta h_k g_k}{6c_k} \tag{6.40}$$
$$\widehat{c}_k = \frac{\Delta h_k}{6} .$$

We can write the matrix equation of residuals (6.39) componentwise and by doing this we have for line k $(k = 1, \ldots, n)$:

— a residual equation $l = N(k - 1) + 1$ for the start point

$$
\begin{aligned}
R_l(t) = & - 2\widehat{c}_k V_l'(t) - \widehat{c}_k V_{l+1}'(t) \\
& - \frac{\widehat{a}_k}{2} V_l(t) - \widehat{b}_k V_{l+1}(t) \\
& - \frac{1}{c_k} \sum_{i=1}^{n} \Big(G_{(2k-1,2i-1)} V_{N(i-1)+1}(t) + G_{(2k-1,2i)} V_{N(i)}(t) \Big) \\
& - \frac{1}{c_k} \sum_{i=1}^{m} G_{(2k-1,2n+i)} V_{N(n)+i}(t) \\
& + \frac{B_{2k-1}(t)}{c_k} \quad ,
\end{aligned}
\tag{6.41}
$$

— a residual equation $l = N(k - 1) + 2, \ldots, N(k) - 1$ for the internal discretization points

$$
\begin{aligned}
R_l(t) = & - \widehat{c}_k V_{l-1}'(t) - 4\widehat{c}_k V_l'(t) - \widehat{c}_k V_{l+1}'(t) \\
& - \widehat{b}_k V_{l-1}(t) - \widehat{a}_k V_l(t) - \widehat{b}_k V_{l+1}(t) \quad \text{and}
\end{aligned}
\tag{6.42}
$$

— a residual equation $l = N(k)$ for the end point

$$
\begin{aligned}
R_l(t) = & - \widehat{c}_k V_{l-1}'(t) - 2\widehat{c}_k V_l'(t) \\
& - \widehat{b}_k V_{l-1}(t) - \frac{\widehat{a}_k}{2} V_l(t) \\
& - \frac{1}{c_k} \sum_{i=1}^{n} \Big(G_{(2k,2i-1)} V_{N(i-1)+1}(t) + G_{(2k,2i)} V_{N(i)}(t) \Big) \\
& - \frac{1}{c_k} \sum_{i=1}^{m} G_{(2k,2n+i)} V_{N(n)+i}(t) \\
& + \frac{B_{2k}(t)}{c_k} \quad .
\end{aligned}
\tag{6.43}
$$

A residual equation $l = N(n + k)$ to the linear capacitor k $(k = 1, \ldots, m)$ reads

$$
\begin{aligned}
R_l(t) = & -s_k V_l'(t) \\
& - \sum_{i=1}^{n} \Big(G_{(2n+k,2i-1)} V_{N(i-1)+1}(t) + G_{(2n+k,2i)} V_{N(i)}(t) \Big) \\
& - \sum_{i=1}^{m} G_{(2n+k,2n+i)} V_{N(n)+i}(t) \\
& + B_{2n+k}(t) \ .
\end{aligned}
\tag{6.44}
$$

6.5. Steady state

The discrete steady state model for (6.1), (6.2) and (6.3) reads

$$\left(\begin{bmatrix} \overline{M} + K & \\ & O^{CAP} \end{bmatrix} + \widetilde{G} \right) V^{\infty} = \widetilde{B}^{\infty} \, , \tag{6.45}$$

where V^{∞} denotes the steady state solution. This has been obtained by leaving out the time derivative terms from the model of the circuit (6.29). We calculate this steady state solution by using subroutine C05PCF from NAG Fortran Library.

6.6 The delay time and its a-priori upper bound

We can calculate the a-priori upper bound for delay time according to relation (5.23). The delay in the discretized model is obtained by the formula

$$\mathbf{D}(t) = \frac{\displaystyle\max_{1 \le k \le N(n)+m} |V_k(t) - V_k^{\infty}|}{\displaystyle\max_{1 \le k \le N(n)+m} |V_k(0) - V_k^{\infty}|} \, . \tag{6.46}$$

The initial condition (6.29)

$$V(0) = (V_1(0), ..., V_{N(n)+m}(0))$$

gives voltages of the FEM model at starting time $t = 0$ and we can see that $0 \le \mathbf{D}(t) \le 1$ for every time t.

Now let us give the formula of the λ-delay time for the FEM model

$$\mathbf{T}_\lambda = \sup_{t \ge 0}\{t; \ \mathbf{D}(t) = \lambda\} \, , \tag{6.47}$$

where $\mathbf{D}(t)$ is given by (6.46). The λ-delay time gives the last time when the difference between the discretized dynamical model and the discretized steady state model has dropped to a λth of the original difference.

6.7 Examples

The purpose of this section is to illustrate the usage of the proposed numerical method for solving some typical mixed circuits. At the same time, we shall numerically compute the delay time given by (6.47) in order to compare it with its a-priori upper bound derived in Chapter V.

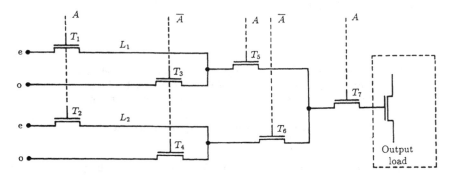

Figure 6.2 NMOS multiplexer

Example 6.1. The circuit shown in Figure 6.2 is a multiplexer realised with NMOS pass transistors.

We want to study the effect of interconnection lines L_1 and L_2 on the signal delay after step sources (with $e = 1$ value) are connected. We suppose that during the transient process the pulses A and $\overline{A}$, applied to the address lines, maintain the transistors T_1, T_2, T_5 and T_7 in the ON state (having the drain–source resistance $R_1 = 1$) and the transistors T_3, T_4, T_6 in the OFF state ($R_2 = 10$). The parameters of the lines are: $r_1 = 1$, $c_1 = 10$, $g_1 = 0.1$, $d_1 = 1$ for the first line L_1 and $r_2 = 1$, $c_2 = 1$, $g_2 = 0.1$, $d_2 = 1$ for the second line L_2. If the output load is modelled by $s = 1$ and $R_3 = 10$, we obtain the mixed type circuit from Figure 6.3:

In Figure 6.4 we present the same circuit with the resistive part grouped in a multiport.

With notations in figure (compatible with the general case in Figure 4.5), Kirchhoff's first and second law give, respectively

$$i_1' + i_2' - j_2 = 0 ; \quad i_3' + i_4' - j_4 = 0 ;$$
$$i_2' + i_3' - i_5' = 0 ; \quad i_5' + i_6' + j_5 = 0 ,$$

and

$$e = R_1 j_1 + w_1 ; \quad 0 = w_2 + R_2 i_1' ; \quad e = R_1 j_3 + w_3 ;$$
$$0 = w_4 + R_2 i_4' ; \quad 0 = w_2 - w_5 + R_1 i_5' + R_1 i_2' ;$$
$$w_5 = R_3 i_6' ; \quad 0 = w_4 - w_5 + R_1 i_5' + R_2 i_3' .$$

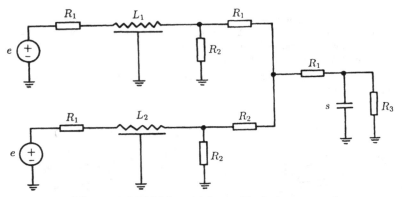

Figure 6.3 NMOS multiplexer. Equivalent network

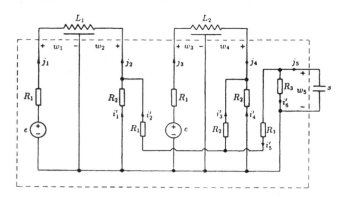

Figure 6.4 Equivalent network with resistive part viewed as a multiport.

By the elimination of $i'_1 - i'_5$ we obtain a linear relation between $j = (j_1, j_2, j_3, j_4, j_5)$ and $w = (w_1, w_2, w_3, w_4, w_5)$ of the form $j = -Gw + B$ where

$$
\begin{bmatrix}
G_1 & 0 & 0 & 0 & 0 \\
0 & \dfrac{G_1^2 + 3G_1 G_2 + G_2^2}{2G_1 + G_2} & 0 & \dfrac{-G_1 G_2}{2G_1 + G_2} & \dfrac{-G_1^2}{2G_1 + G_2} \\
0 & 0 & G_1 & 0 & 0 \\
0 & \dfrac{-G_1 G_2}{2G_1 + G_2} & 0 & \dfrac{4G_1 G_2 + G_2^2}{2G_1 + G_2} & \dfrac{-G_1 G_2}{2G_1 + G_2} \\
0 & \dfrac{-G_1^2}{2G_1 + G_2} & 0 & \dfrac{-G_1 G_2}{2G_1 + G_2} & \dfrac{G_1^2 + G_1 G_2 + 2G_1 G_3 + G_2 G_3}{2G_1 + G_2}
\end{bmatrix}
$$

and $B = (eG_1, 0, eG_1, 0, 0)^{tr}$. Here we have put $G_i = 1/R_i$. The initial values are all zero: $v_{1,0}(x) = v_{2,0}(x) = v_{3,0} = 0$. We choose for the two lines the same space discretization step: $\Delta h_1 = \Delta h_2 = d_1/20 = d_2/20 = 0.05$, while for the 273 time steps the minimum and maximum time steps are 2.155×10^{-6} and 3.547 respectively, in NAG subroutine D02NGF.

The Figs 6.5, 6.6 and 6.7 show the voltage variation along lines and on the capacitor, respectively.

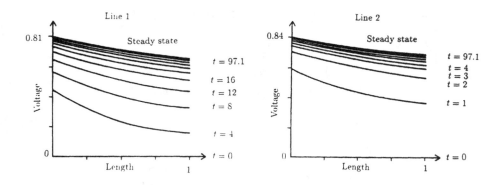

Figure 6.5 $v_1(t,x)$ **Figure 6.6** $v_2(t,x)$

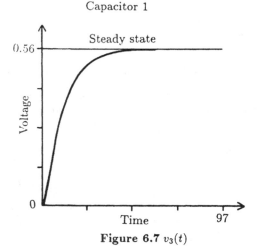

Figure 6.7 $v_3(t)$

As we see, all components of the solution tend to the steady state. Because G has the DRD property, this is the confirmation of the global stability result from Theorem 5.1 ii). The speed of the evolution to the steady state differs from one element to the other. So, from Figures 6.5 and 6.6 we observe that the voltage along the first line varies (approximately) ten times slower than the second line voltage. That is in accordance with the "engineering feeling", because $R_1C_1 = 10R_2C_2$, where $R_1 = r_1d_1$, $R_2 = r_2d_2$, $C_1 = c_1d_1$, $C_2 = c_2d_2$ are the "lumped equivalent" resistances and capacitances of the lines.

A good image of the global speed of evolution of the circuit is given by the delay $D(t)$ from Fig 6.8. This gives the delay time values: $T_{0.5} = 6.11$ s and $T_{0.1} = 21.3$ s.

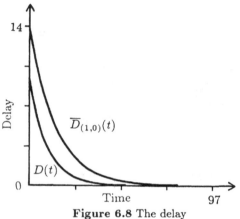

Figure 6.8 The delay

In the same picture we see the upper bound $\overline{D}_{(1,0)}(t)$ of the delay. For its calculation, the program uses the formulae (5.7), (5.21) and (5.22) with the intermediate results from Table 6.1.

σ	$\cos\gamma_1^\sigma$	$\cos\gamma_2^\sigma$	ω_0^σ	$\overline{T}_{0.5}^\sigma$	$\overline{T}_{0.1}^\sigma$
(0,0)	0.962	0.955	-0.0175	42.1	134.0
(1,0)	0.707	0.955	-0.717	14.0	37.0
(0,1)	0.962	0.707	-0.0175	59.2	151.0
(1,1)	0.707	0.707	-0.0716	14.5	37.0

Table 6.1

According to (5.23) we find $\overline{T}_{0.5} = 14.5$ and $\overline{T}_{0.1} = 37.0$, for $\sigma = (1,0)$, which are very close to the "exact" values $T_{0.5}$ and $T_{0.1}$ from above.

Example 6.2. Another usual kind of network, composed of conducting paths between active devices in digital integrated circuits, is the so called "tree-type" network, Rubinstein et al. [1983]. This is defined as a circuit in which there exists a unique path connecting a specified input with a specified output. Here by "input" we mean a given inverter or a logic node, while an "output" is a gate driven by that input. For example in Figure 6.9 the pairs (E_1, R_1) and (E_2, R_1) model two inputs and the groups (R_1, s_1), (R_1, R_2, s_2), (R_1, R_2, s_3) and (R_1, R_2, s_4) are possible models for outputs.

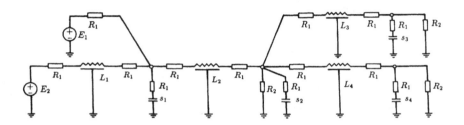

Figure 6.9 A tree network

The voltage values of inputs are $E_1 = E_2 = 1$ and the lumped elements have the values $R_1 = 1/2$, $R_2 = 1$, $s_1 = 3$, $s_2 = 3/2$, $s_3 = 3$ and $s_4 = 3$. The four rcg–lines have the same length $d = 1$, the same resistance (per unit length) $r = 1$, the same conductance $g = 0$ and $c_1 = 3/2$, $c_2 = 1$, $c_3 = 5/4$, $c_4 = 4/3$. All dimensions are coherent. This results in a 12×12 symmetric G matrix whose diagonal elements are $G_{11} = 2$, $G_{22} = G_{33} = G_{99} = 3/2$, $G_{44} = G_{55} = G_{77} = G_{10,10} = 14/9$, $G_{66} = G_{88} = G_{11,11} = G_{12,12} = 6/5$ and the nondiagonal, nonzero ones are $G_{23} = G_{32} = G_{29} = G_{92} = G_{39} = G_{93} = -1/2$, $G_{45} = G_{54} = G_{47} = G_{74} = G_{4,10} = G_{10,4} = G_{57} = G_{75} = G_{5,10} = G_{10,5} = G_{7,10} = G_{10,7} = -4/9$, $G_{6,12} = G_{12,6} = G_{8,11} = G_{11,8} = -4/5$. The nonzero elements of B are $B_1 = 2$ and $B_2 = B_3 = B_9 = 1/2$. The initial state is zero for all lines and all capacitors.

The choosen space step is $\Delta h = d/10$ and the 229 time steps vary between 5.75×10^{-6} and 9.37×10^{-1}. We shall reproduce here, in pictures, only some of the results. Figures 6.10 and 6.11 show the voltage along the first and the fourth lines respectively for different moments.

Analogously, Figures 6.12 and 6.13 show the voltage evolution for the first and the fourth capacitor, respectively.

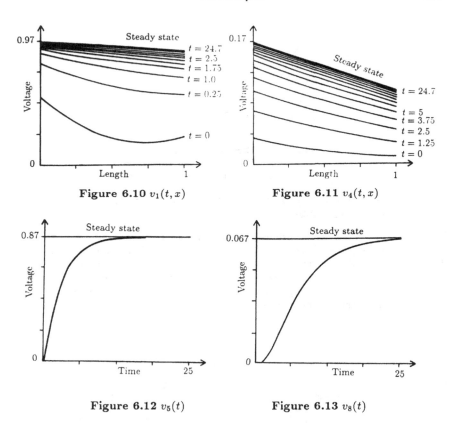

Figure 6.10 $v_1(t,x)$ **Figure 6.11** $v_4(t,x)$

Figure 6.12 $v_5(t)$ **Figure 6.13** $v_8(t)$

First of all, we remark that all state values advance to a steady state, as our theoretical result predicted. Then we find that it is a process of propagation of signals from inputs to outputs. So, while for the line 1 (which is near to inputs) the steady state has the values $v_{1,\infty}(0) = 0.966$ and $v_{1,\infty}(d) = 0.899$, for the line 4 (which is close to output) these values are $v_{4,\infty}(0) = 0.166$ and $v_{4,\infty}(d) = 0.099$. The same happens with capacitors s_1 and s_4 (0.866 compared to 0.066). The above pictures also clearly show that the elements which are close to sources (inputs) tend to the steady state more quickly than those that are far away, near to outputs. The global switching speed of the circuit is expressed by the moment when the delay $D(t)$ (see Figure 6.14) pass through an imposed λ value. This is the λ–delay time and in our case the numerical integration gives $T_{0.5} = 1.96$ and $T_{0.1} = 6.38$.

With regards the upper bound of the delay time, this was found for the best

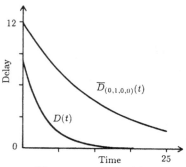

Figure 6.14 The delay

permutation $\sigma = (0, 1, 0, 0)$, giving $\overline{T}_{0.5} = 9.3$ and $\overline{T}_{0.1} = 18.4$. Again the tightness of these bounds is reasonable.

Example 6.3. Now let us consider a mesh network. This kind of network arises in models for the gates of large transistors (used in the final stages of clock or pad drivers) or for CMOS transmission gates, Wyatt [1985]. The Manchester carry adder with carry–bypass circuitry, Chan and Schlag [1989], Chan and Karplus [1990], provides another example of a digital network containing closed loops. Let us consider the (very symmetrical) network in Figure 6.15.

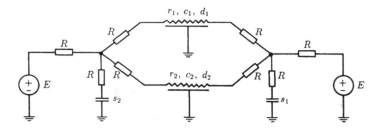

Figure 6.15 A simple mesh network

The distributed parameters of the lines are $r_1 = 10$, $c_1 = 10$, $g_1 = 0$, $r_2 = 30$, $c_2 = 30$, $g_2 = 0$, the length being the same $d_1 = d_2 = 1$. The capacitances are $s_1 = 100$ and $s_2 = 200$, the resistance is $R = 1$ and the two step sources (simultaneous connected) have the voltage $E = 1$. The initial values are zero.

The input data for the numerical program comprises of the matrix G with elements: $G_{ii} = 3/4$, $i = 1, ..., 6$, $G_{13} = G_{31} = G_{16} = G_{61} = G_{24} = G_{42} = G_{25} =$

$G_{52} = G_{36} = G_{63} = G_{45} = G_{54} = -1/4$ and the others being zero. All six elements of vector B are 1/4. We have used the discretization space step $\Delta h_i = d_i/20$, $i = 1, 2$, while the minimum and the maximum time steps in the NAG subroutine D02NGF are 1.049×10^{-4} and 2 respectively, the number of time steps being 1515.

The numerical results, graphically presented in Figures 6.16–6.19, confirm the good work of our program. This is because, the symmetry of the circuit a-priori shows that the steady state must have the same voltage =1 both on capacitors and on all points of the two lines.

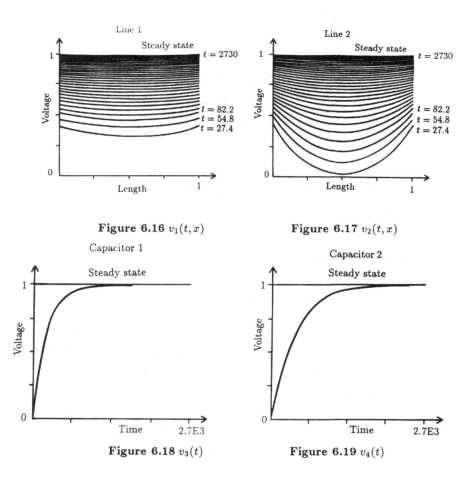

Figure 6.16 $v_1(t, x)$ **Figure 6.17** $v_2(t, x)$

Figure 6.18 $v_3(t)$ **Figure 6.19** $v_4(t)$

Also, the speed of the transient regime differs from one element to the other according to engineering intuition. The global delay is shown in Figure 6.20 together with its a-priori upper bound.

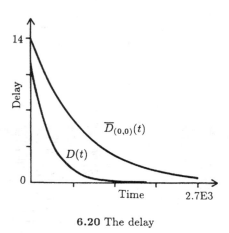

6.20 The delay

This second one is computed with the "best" permutation $\sigma = (0, 0)$ giving $\omega_0^\sigma = -0.705 \times 10^{-3}$ and $\min_{i=1,2} \cos \gamma_i^\sigma = 0.6991$. As a consequence we find $\overline{T}_{0.5} = 1490$ and $\overline{T}_{0.1} = 3770$ as upper bounds for the "exact" delay times $T_{0.5} = 295$ and $T_{0.1} = 935$, respectively.

6.8. Concluding remarks

Let us underline here the main applicative aspects of the models and methods presented in Chapters IV, V and VI.

The problem of delay time prediction is one of the crucial aspects in digital circuit design. In modern technology the improvement of this performance becomes mainly limited by interconnection parasitics. This is why we have introduced an accurate and general model for a network in which connecting wires are modelled more exactly as distributed parameter elements while the devices are lumped approximated. To use such a "mixed type" mathematical model, care should be taken in obtaining the conditions for its validity. As we have seen, our conditions are in good agreement with the engineering necessities, assuming the large applicativity of our new model. The second problem solved above was to derive an upper bound of the delay time for our mixed-type circuit. The formula giving this bound is simple and therefore appropriate for engineering use in design. The third problem about our mixed-type model, was its numerical realization and implementation. The discretization was based on variational formulation of the problem and on the use of the finite element method in space discretization. By this method it is easy and natural to handle the "crossed" boundary conditions. The semidiscrete model (called FEM-model above) leads to an initial value problem for systems of differential equations. Typically this system is stiff. As described in Appendix I, one can combine the FEM-model with a general purpose subroutine library such as NAG. Otherwise, such a sophisticated subroutine library makes it possible to build up a fairly general electrical simulator including nonlinear capacitors and even lumped nonlinearly modelled transistors (not described here). Having our circuit simulator built, we have verified our delay time bound for some typical examples. A reasonable tightness was found.

Appendix I

The purpose of this Appendix is to outline how to use subroutine D02NGF of the NAG Fortran Library in solving the initial value problem (6.29_1), (6.29_2). As explained in Chapter 6.4, we shall apply integrators developed for stiff systems of implicitely defined ordinary differential equations

$$A(t,y)y' = g(t,y) \quad . \tag{A1}$$

This formulation permits the solution of differential/algebraic systems (DAEs). One evaluates the residual vector

$$r(t,y) = g(t,y) - A(t,y)y' \tag{A2}$$

to get solution y at time t.

The general form of a program calling D02NGF is (see Figures A2–A3)

```
declarations
    :
call linear algebra setup routine (D02NSF)
call integrator setup routine (D02NVF or D02NWF)
call integrator (D02NGF)
call integrator diagnostic routine (D02NYF)
    :
END
```

We set up the full matrix linear algebra (D02NSF) and choose either Backward Differentiation formulae (D02NVF) or BLEND formulae (D02NWF) as the integration method with adaptive time stepping. The numerical results in Chapter 6.6 have been obtained by using Backward Differentiation Formulae because it copes fewer iterations than BLEND formulae. The integration routine D02NGF calls user-defined subroutines RESID to calculate $r(t,y)$, JAC to calculate the Jacobian of the residual system and MONITR to cope with error situations and to change some integration parameters. The form of those residuals is given by formulae

(6.41)–(6.44). The (i,j)th component of the Jacobian matrix $\dfrac{\partial r(t,y)}{\partial y}$ of the residual system must be given in the form

$$\frac{\partial r_i(t,y)}{\partial y_j} = -A_{ij} + hd\frac{\partial g_i(t,y)}{\partial y_j} \ , \tag{A3}$$

where h is the current stepsize and d is a parameter depending on the integration method. Diagnostic routine D02NYF gives us some values concerning integration. For further information on how to use the subroutines see NAG user's manual, [NAG].

In Figure A1 we can see the way to build up the semidiscrete system from the system of partial differential equations (6.1) and boundary conditions given in the form of differential equations (see (6.2), (6.6) and (6.7)). This is done by using the finite element method in discretization of space variables of rcg-lines. Capacitors are added to the FEM-model of lines by (6.8). In Figures A2–A3 we have outlined the time integration of the semidiscrete system in the developed program. The program also gives us delay and a-priori upper bound of delay at time moments with equal intervals. The values of parameters needed in the calculation of the solution and the upper bound at every time discretization step are calculated in the initialization part of the program by using circuit parameters. The calculation of delay is done at every time step by using formula (6.46).

BUILDING THE SEMIDISCRETE SYSTEM OF THE CIRCUIT.

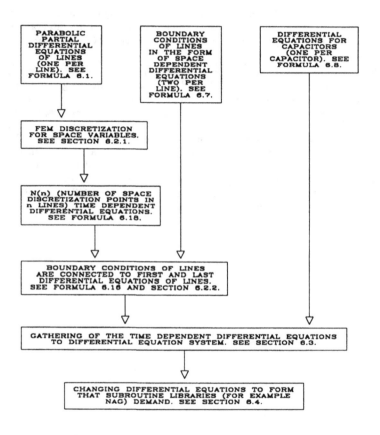

Figure A1

THE INITIALIZATION
OF THE PARAMETERS.

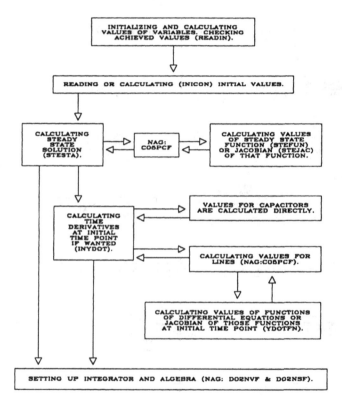

Figure A2

STRUCTURE OF PROGRAM CIRCUITS

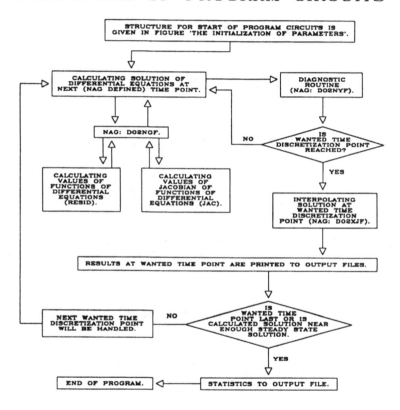

Figure A3

List of symbols

$\mathbf{R}^n$	n-dimensional Euclidean space
$\mathbf{C}$	set of complex numbers
$\mathbf{N}$	set of natural numbers
$\mathbf{Z}$	set of integer numbers
$]a, b[$	open interval
$[a, b]$	closed interval
$\partial u/\partial x_i$	partial derivative of u with respect to x_i
$\partial\|z\|$	subdifferential
$d^- f(t)/dt \quad (d^+ f(t)/dt)$	left-hand (right-hand) derivative
$\ln$	natural logarithm
$\det B$	determinant of matrix B
B^{tr}	transposed matrix
$\operatorname{Re} f$	real part of f
δ_{ij}	Kronecker's symbol: $\delta_{ij} = 1$ for $i = j$, otherwise $\delta_{ij} = 0$
$B(x, r)$	$\{y \in X \; ; \; \|y - x\| < r\}$ open ball with radial r and central point x
$S(x, r) = \partial B(x, r)$	$\{y \in X \; ; \; \|y - x\| = r\}$ boundary of $B(x, r)$
X^*	dual space of X
$\xrightarrow{w}$	weak convergence
$\|x\|$	norm on Banach space X
$\|x\|_d$	weighted norm
$\{p_f(x) = f(x) \; ; \; f \in X^*\}$	seminorm
$\langle,\rangle_+, \langle,\rangle$	"duality functionals"
$\|f\| = \sup\limits_{\|x\|<1} \|f(x)\|$	norm in dual space X^*
$\|f\| = \max\limits_{1\leq i\leq n} \max\limits_{x\in Y} \|f_i(x)\|$	supremum norm
$\mathcal{D}(A)$	domain of definition for operator A
$\mathcal{R}(A)$	range of operator A
$\rho(A)$	resolvent set of operator A
$A \oplus B$	$\begin{bmatrix} A & 0 \\ 0 & B \end{bmatrix}$
$\mu_d(M)$	measure of $N \times N$ matrix M
ℓ^p	Banach space of all complex functions x on $\{0, 1, 2, ...\}$ whose norm $\|x\|_p = \{\sum\limits_{k=0}^{\infty} |x(k)|^p\}^{1/p}$ is finite
$\mathbf{K}_s^m$	the space $\mathbf{K}^m$ ($\mathbf{K} = \mathbf{R}$ or $\mathbf{C}$) with euclidean norm weighted by positive constants $s_1, ..., s_m$

$C(Y, \mathbf{R}^n)$	$\{f : Y \mapsto \mathbf{R}$ with continuous f_i components $\}$ with the supremum norm
$C^m(0, T; X)$	a space of functions $u :]0, T[\mapsto X$ with continuous derivatives up to order m
$C_0^\infty(0, d_i; \mathbf{K})$	a space of functions from $C^\infty(0, d_i; \mathbf{K})$ with compact support
$L_1(0, T; X)$	a space of integrable functions
$L^p(Y), p < \infty$	Lebesgue space of measurable functions v defined on Y for which $[\int_Y |v(x)|^p \, dx]^{1/p}$ is finite
$L_{2,i}(0, d_i; \mathbf{K})$	weighted L^2-space with norm $\|f\|_{L_{2,i}} = \int_0^{d_i} c_i |f(x)|^2 \, dx$
L_2^n	$= \prod_{i=1}^n L_{2,i}$
$H_{m,i}$	$\{f \in L_{2,i} \mid \frac{\partial^\alpha}{\partial^\alpha x_i} f \in L_{2,i}, |\alpha| \leq m\}$
H_m^n	$= \prod_{i=1}^n H_{m,i}$
$X_\mathbf{K}$	$= L_2^n(\mathbf{K}) \times \mathbf{K}_s^m$
$Y_\mathbf{K}$	$= H_1^n(\mathbf{K}) \times \mathbf{K}_s^m$
$L^\infty(Y)$	Lebesgue space of measurable essentially bounded functions defined on Y
$C(X)$	a space of continuous functions defined on X with norm $\| \cdot \|_X$
$C^k(]0, T[; X)$	a space of all continuous functions $u : [0, T] \mapsto X$ with norm $\| \cdot \|_X$
$C^{1+\nu}(0, T; X)$	$\{f \in C^1(0, T; X) ; \frac{d}{dt} f \in C^\nu(0, T; X)\}$
C_E^p	$\{f : E \mapsto \ell^p ; f$ is continuous$\}$
$c, \hat{c}, c_i, ...$	generic constant
$\exists$	there exist(s)
$\forall$	for all
a.e.	almost everywhere
v	voltage
i	current
r	resistance
c	capacitance
g	conductance
G	matrix of boundary conditions
B	source vector
$(G, B, 2n + m)$	resistive multiport
$D(t)$	delay
T_λ	$= \sup\{t \mid D(t) = \lambda\}$ λ delay time

Index

References

Agmon, S., "Lectures on elliptic boundary value problems," Van Nostrand, New York, 1965.

Araki, K. and Naito, Y., *Computer-aided analysis of coupled lossy transmission lines*, in Proc. Intern. Symp. on Circ. and Syst., 423–426, 1985.

Bakoglu, H.B. and Meindl, J.D., *Optimal interconnection circuits for VLSI*, IEEE Trans.Electr.Der. **ED 32 No 5**, 903–909, 1985.

Barbu, V., "Nonlinear Semigroups and Differential Equations in Banach Spaces," Noordhoff, Leyden, 1976.

Barbu, V., *Nonlinear boundary-value problems for a class of hyperbolic systems*, Rev. Roum. Math. Pures et Appl. **22**, 155–168, 1977.

Barbu, V. and Morosanu, G., *Existence for a nonlinear hyperbolic system*, Nonlinear Anal. T.M.A. **5**, 341–353, 1981.

Berzins, M., Dew, P.M. and Furzeland, R.M., *Developing software for time-dependent problems using the method of lines and differential-algebraic integrators*, Appl. Numer. Math. **5** (1989), 375–397.

Bose, M.P. and Showalter, R.E., *Homogenization of the layered medium equation. Appl. Anal.*, 1990.

Brayton, R.K., *Small signal stability criterion for electrical networks containing lossless transmission lines*, IBM J. Research Dev. **12**, 431–440, 1968.

Brayton, R.K. and Miranker, W.L., *A stability theory for nonlinear mixed initial boundary value-problems*, Arhch. Rational Mech. Anal. **17**, 358–376, 1964.

Brézis, H., "Operateurs Maximaux Monotones et Semigroupes de Contractions dans les Espáces de Hilbert," North-Holland, Amsterdam, 1973.

Brézis, H., "Analyse Fonctionelle. Théorie et Applications," Masson, Paris, 1987.

Browder, F.E., *Nonlinear accretive operators in Banach spaces*, Bull. Amer. Math. Soc. **73**, 470–475, 1967.

Butzer, P.L. and Berens, H., "Semi-Groups of Operators and Approximation," Springer-Verlag, Berlin, 1967.

Cases, M. and Quinn, D.M., *Transient response of uniformly distributed RLC transmission lines*, IEEE Trans.Circ.Syst. **27**, 200–207, 1980.

Chan, P.K., *An extension of Elmore's delay*, IEEE Trans. Circ. Syst., CAS **33 No 11**, 1147–1149, 1986 a.

Chan, P.K., *An extension of Elmore's delay and its application for timing analysis of MOS pass transistor networks*, IEEE Trans. Circ. Syst., Cas **33 No 11**, 1149–1152, 1986 b.

Chan, P.K. and Karplus, K., *Computing signal delay in general RC networks by tree/link partitioning*, IEEE Trans.Comp.Aided Design **CAD 9 No 8**, 898–902, 1990.

Chan, P.K. and Schlag, M.D.F., *Bounds on signal delay in RC mesh networks*, IEEE Trans. Circ. Syst. **CAS 8 No 6**, 581–589, 1989.

Chew, K.W., Shivakumar, P.N. and Williams, J.J., *Error bounds for the truncation of infinite linear differential systems*, J. Inst. Math. Applic. **25**, 37–51, 1980.

Chien, M.J., *Piecewise-linear theory and computation of solutions of homeomorphic resistive networks*, IEEE Trans. Circ. Syst. **CAS 24**, 118–127, 1977.

Chua, L.O. and Green, D.N., *A qualitative analysis of the behaviour of dynamic nonlinear networks – stability of autonomous networks*, IEEE Trans. Circ. Syst., CAS **23 No 6**, 355–379, 1976 a.

Chua, L.O. and Green, D.N., *A qualitative analysis of the behaviour of dynamic nonlinear networks: steady-state solutions of nonautonomous networks*, IEEE Trans. Circ. Syst., CAS **23 No 9**, 530–550, 1976 b.

Ciesielsni, M.J., *Layer assignment for VLSI interconnect delay minimization*, IEEE Trans.Comp. Aided Design **CAD 8 No 6**, 702–707, 1989.

Cooke, K.L. and Krumme, D.W., *Differential-Difference equations and nonlinear initial boundary value problems for linear hyperbolic partial differential eqations*, J. Math. Anal. Appl. **24**, 372–387, 1968.

Coppel, W.A., "Stability and Asymptotic Behavior of Differential Equations," D.C.Heath, Boston, 1965.

Deimling, K., "Ordinary Differential Equations in Banach Spaces," Lecture Notes in Math., **596**, Springer-Verlag, Berlin, 1977.

Desoer, C.A., *Distributed networks with small parasitic elements: input-output stability*, IEEE Trans. Circ. Syst., CAS **24**, 1–7, 1977.

Desoer, C.A. and Haneda, H., *The measure of a matrix as a tool to analyze computer algorithms for circuit analysis*, IEEE Trans. Circ. Syst., CT **19 No 5**, 480–486, 1972.

Desoer, C.A. and Katzenelson, J., *Nonlinear RLC networks*, Bell Syst. Tech. J. **44 No 1**, 161–198, 1965.

Desoer, C.A. and Kuh, E.S., "Basic circuit theory," McGraw–Hill, New York, 1969.

Dolezal, V., "Nonlinear networks," Elsevier Sci. Publ. Co., New York, 1977.

Dolezal, V., "Monotone operators and applications in control and network theory," Elsevier Sci. Publ. Co., New York, 1979.

Dunford, N. and Schwartz, J.T., "Linear operators, Part I," J. Wiley, Intersience Publ., New York, 1958.

Elmore, W.C., *The transient response of damped linear networks with particular regard to wide-band amplifiers*, J. Appl. Phys. **19**, 55–63, 1948.

Fattorini, H.O., "The Cauchy Problem," Addison-Wesley, Reading, 1983.

Fujisawa, T. and Kuh, E.S., *Some results on existence and uniqueness of solutions of nonlinear networks*, IEEE Trans. Circ. Th. **CT–18**, 389–394, 1971.

Gao, D.S., Yang, A.T. and Kang, S.M., *Modelling and simulation of interconnection delays and crosstalks in high-speed integrated circuits*, IEEE Trans.Circ.Syst. **CAS 37**, 1–9, 1990.

Ghausi, M.S. and Kelly, J.J., "Introduction to Distributed Parameter Networks," Holt, Rinehart and Winston Inc., New York, 1968.

Glasser, L.A. and Dobberphul, D.W., "Design and Analysis of VLSI Circuits," Addison-Wesley, New York, 1985.

Glez. Harbour, M. and Drake, J.M., *Simple RC model for integrated multiterminal interconnections*, IEE Proceedings, Part G, **135, No 1**, 19–23, 1988.

Goldstein, J.A., "Semigroups of operators and abstract Cauchy problems," Lecture notes, Tulane university, 1970.

Grondin, R.O., Porod, W. and Ferry, D.K., *Delay time and signal prapagation in large-scale integrated circuits*, IEEE J. Solid-State Circ., SC **19 No 2**, 262–263, 1984.

Gummel, H.K., *A charge-control transistor model for network analysis programs*, Proc.IEEE **56**, 751–752, 1968.

Halanay, A., "Differential Equations: Stability, Oscilations, Time Lags," Acad. Press, New York, 1965.

Hamilton, D.J.M. and Howard, W.G., "Basic Integrated Circuit Engineering," McGraw-Hill, New York, 1975.

Hamilton, D.J. Lindholm, F.A. and Marshak, A.H., "Principles and Applications of Semiconductor Device Modelling," Holt, Rinehart and Winston Inc., New York, 1971.

Harbour, M.G. and Drake, J.M., *Calculation of signal delay in integrated interconnections*, IEEE Trans. Circ. Syst. CAS **36, No 2**, 272–276, 1989.

Haslinger, J. and Neittaanmäki, P., "Finite element approximation of optimal shape design. Theory and Applications," J. Wiley & Sons, Chichester, 1988.

Hellwig, G., "Differential Operators of Mathematical Physics," Addison-Wesley, Reading, 1967.

Jain, N.K., Prasad, V.C. and Bhattacharyya, A.B., *Delay time sensitivity in linear RC tree*, IEEE Trans. Circ. Syst., CAS **34 No 4**, 443–445, 1987.

Jouppi, N.P., *Timing analysis and performance improvement of MOS-VLSI design*, IEEE Trans. Comp. Aided Design, CAD **6 No 4**, 650–665, 1987.

Kato, T., *Nonlinear semigroups and evolution equations*, J. Math. Soc. Japan **19**, 509–520, 1967.

Kenmochi, N. and Takahashi, T., *On the global existence of solutions of differential equations on closed subsets of a Banach space*, Proc. Japan Acad. Ser. A Math. Sci. **51**, 520–525, 1975.

Kim, Y.H., "ELOGIC: A relaxation-based switch-level simulation technique," UCB/ERL M86/2, Univ of California, Berkley, 1986.

Kolmogorov, A. and Fomine, S., "Elements of the theory of functions and functional analysis," Graylock Press, Baltimore, 1957.

Křižek, M and Neittaanmäki, P., "Introduction to finite element method for variational problems with applications," Pitman Monographs in Pure and Applied Mathematics 50, Longman, Essex, 1990.

Kumar, U., *A bibliography of distributed RC networks*, IEEE Circ. Syst. Mag. **2 No 2**, 9–19, 1980.

Lin, T.M. and Mead, C.A., *Signal delay in general RC networks*, IEEE Trans. Comp. Aided Design. CAD **3 No 4**, 331–349, 1984.

Lin, T.M. and Mead, C.A., *A hierarchical timing simulation model*, IEEE Trans. Comp. Aided Design. CAD **5 No 1**, 188–197, 1986.

Lions, J.L. and Magenes, E., "Non-homogenous boundary value problems and applications," Vol II, Springer-Verlag, Berlin, 1972.

Lovelady, D.L. and Martin, R.H.Jr., *A global existence theorem for a nonautonomous differential equation in a Banach space*, Proc. Amer. Math. Soc. **35**, 445–449, 1972.

Lumer, G. and Phillips, R.S., *Dissipative operators in Banach space*, Pacific J.Math II, 679–698, 1961.

Lusternik, L.A. and Sobolev, V.J., "Elements of functional analysis," Hindustan Publ. Corp. John Wiley, Delhi-New York, 1974.

Marinov, C.A., *Qualitative properties of ℓ^p-solutions of infinite differential systems via dissipativity*, Nonlinear Anal., T.M.A. **8**, 441–456, 1984.

Marinov, C.A., *Truncation errors for infinite linear systems*, IMA J .Num. Anal. **6**, 51–63, 1986.

Marinov, C.A., *The delay time for a rcg line*, Int. J. Circ. Th. Appl. **15**, 79–83, 1987.

Marinov, C.A., *Sandberg type properties for nonlinear transistor networks*, Proc. Eur. Conf. Circ. Th. Design, 53–58, Paris, 1987.

Marinov, C.A., *On assymtotic stability of nonlinear networks*, IEEE Trans. Circ. Syst. **CAS 37, No 4,**, 571–573, 1990 a.

Marinov, C.A., *Dissipativity as an unified approach to Sandberg-Wilson type properties of nonlinear transistor networks*, Int. J. Circ. Th. Appl. **18, No 6**, 575–594, 1990 b,.

Marinov, C.A. and Lehtonen, A., *Mixed-type circuits with distributed and lumped parameters*, IEEE Trans. Circ. Syst. **CAS 36 No 8**, 1080–1086, 1989.

Marinov, C.A. and Moroşanu, G., *Consistent models for electrical networks with distributed parameters*, Časopis pro pěstování matematiky, to appear 1991.

Marinov, C.A. and Neittaanmäki, P., *Delay time for general disrtibuted networks with applications to timing analysis of digital MOS integrated circuits*, in "Simulation of Semiconductor Devices and Processes," K. Board (ed.), Pineridge Press, Swansea, 322–326, 1986.

Marinov, C.A. and Neittaanmäki, P., *A theory of electrical circuits with resistively coupled distributed structures. Delay time predicting*, IEEE Trans. Circ. Syst. **CAS 35 No 2**, 173–183, 1988.

Marinov, C.A. and Neittaanmäki, P., *Global delay time for general distributed networks with applications to timing analysis of digital MOS integrated circuits*, Int.J.Comp.Math.Electrical Electronic Eng., COMPEL **8 No 1**, 17–37, 1989.

Marinov, C.A. and Neittaanmäki, P., *Asymptotical convergence evaluation for a parabolic problem arising in circuit theory*, Z. Angew. Math. Mech. **70 No 8**, 344–347, 1990 a.

Marinov, C.A. and Neittaanmäki, P., *A delay time bound for disrtibuted parameter circuits with bipolar transistors*, Int. J. Circ. Th. Appl., **18**, 99–106, 1990 b.

Marinov, C.A. and Neittaanmäki, P., *Bounds for the solution of a system of parabolic equations arising in circuit theory*, 13th IMACS World Congress on Comp. Appl. Math., Dublin, 1991 a.

Marinov, C.A. and Neittaanmäki, P., *Both sided estimates for distributed structures arising in MOS interconnections*, Eur. Conf. Circ. Th. Design, ECCTD, Copenhagen, 1991 b.

Marinov, C.A., Neittaanmäki, P. and Hara, V., *Signal delay in general distributed networks*, Proc. XXX Int. Conf. ETAN **3**, 35–43, Herceg-Novi, 1986.

Marinov, C.A., Neittaanmäki, P. and Hara, V., *Numerical approach for signal delay in general distributed networks*, Proc. Vth Int. Conf. Num. Anal.

Semicond. Dev. Integr. Circ., J.J.H. Miller ed., 307–312, Boole Press, Dublin, 1987.

Marinov, C.A., Neittaanmäki, P. and Hara, V., *A consistent model for the wiring delay of the MOS inverter*, Proc, ECCTD, 89–93, Brighton, 1989.

Martin, R.H.Jr., *The logarithmic derivate and equations of evolutions in a Banach space*, J. Math. Soc. Japan **22**, 411–429, 1970 a.

Martin, R.H.Jr., *A global existence theorem for autonomous differential equations in a Banach space*, Proc. Amer. Math. Soc. **26**, 307–314, 1970 b.

Martin, R.H.Jr., "Nonlinear Operators and Differential Equations in Banach Spaces," John Wiley, New York, 1976.

McClure, J.P. and Wong, *On infinite systems of linear differential equations*, Can. J. Math. **27**, 691–703, 1975.

McClure, J.P. and Wong, *Infinite systems of differential equations*, Can. J. Math. **28**, 1132–1145, 1976.

McClure, J.P. and Wong, *Infinite systems of differential equations II*, Can. J. Math. **31**, 596–603, 1979.

Mikhailov, V., "Partial differential equation," Mir, Moscow, 1978.

Miller, R.K. and Michel, A.N., *Stability theory for countable infinite systems of differential equations*, Tohoku Math. J. **32**, 155–168, 1980.

Minty, G.J., *Monotone networks*, Proc. Royal Soc. (London), Ser A **257**, 194–212, 1960.

Minty, G.J., *Solving steady-state nonlinear networks of monotone elements*, Trans. Circ. Th, CT **8**, 99–104, 1961.

Moroşanu, G., *Stability of solutions of nonlinear boundary-value problems for hyperbolic systems*, Nonlinear Anal., TMA **5**, 61–70, 1981, a.

Moroşanu, G., *Mixed problems for a class of nonlinear differential hyperbolic systems*, J. Math. Anal. Appl. **83**, 470–485, 1981, b.

Moroşanu, G., *On a class of nonlinear differential hyperbolic systems with non-local boundary conditions*, J. Diff. Eq. **43**, 345–368, 1982.

Moroşanu, G., "Nonlinear Evolution Equations and Applications," D. Reidel Publ. Co., Dordrecht, 1988.

Moroşanu, G. and Petrovanu, D., *Nonlinear monotone boundary conditions for parabolic equations*, Rend. Ist. Matem. Univ. Trieste **18**, 136–155, 1986.

Moroşanu, G., Marinov, C.A. and Neittaanmäki, P., *Well-posed nonlinear problems in the theory of electrical networks with distributed and lumped parameters*, Proc. of Int. Conf. Simul. Modelling, IASTED, Lugano,, 345–348, 1989.

Moroşanu, G., Marinov, C.A. and Neittaanmäki, P., *Well-posed nonlinear problems in integrated circuits modelling*, Circ. Syst. Sign. Proc., 10, 53–69, 1991.

Nagel L.W., *SPIECE 2: A computer program to simulate semiconductor circuits*, Univ. of California, Berkeley, ERL Memo, No ERL-M250, 1975.

Neittaanmäki, P., Hara, V. and Marinov, C.A., *Numerical approach for signal delay in general distributed networks*, Proc., IEEE Int. Symp. Circ. Syst. **2**, 1353–1358, Helsinki, 1988.

Nakhla, M.S., *Analysis of pulse propagation on high–speed VLSI chips*, IEEE J.Solid State Circ. **25 No 2**, 490–494, 1990.

O'Brien, P. and Wyatt, J.L. Jr., *Signal delay in ECL interconnect*, in Proc. IEEE Int. Symp. Circ. Syst., 1986.

Ôharu, S., *Note on the representation of semi-groups of nonlinear operators*, Proc. Jap. Acad. **XLII**, 1966.

Ortega, J.M. and Rheinboldt, W.C., "Iterative solutions of nonlinear equations in several variables," Academic Press, New York, 1970.

Ousterhout, J.K., *A switch-level timing verifier for digital MOS VLSI*, IEEE Trans. Comp. Aided Design, CAD **4 No 7**, 336–349, 1985.

Pao, C.V., *On the asymptotic stability of differential eqations in Banach spaces*, Math. Syst. Th. **7**, 25–31, 1973.

Passlack, M., Uhle, M. and Elschner, H., *Analysis of propagation delays in high–speed VLSI circuits using a distributed line model*, IEEE Trans.Comp. Aided Design **CAD 9 No 8**, 821–826, 1990.

Pavel, N.H., *Sur certaines équations différentielles abstraites*, Boll. Un. Mat. Ital. **6**, 397–409, 1972 a.

Pavel, N.H., *Sur certaines équations différentielles non-linéaires dans un espace de Banach*, C. R. Acad. Sci. Paris Sér.A **275**, 1183–1185, 1972 b.

Pavel, N.H., "Differential equations, flow invariance and applications," Pitman Research Notes in Math. 113, Boston, 1984.

Pazy, A., "Semigroups of Linear Operators and Applications to Partial Differential Equations," Appl.Math.Sci., Springer-Verlag, New York, 1983.

Phillips, R.S., *Dissipative hyperbolic systems*, Trans. Amer. Math. Soc. **86**, 109–173, 1957.

Phillips, R.S., *Dissipative operators and hyperbolic systems of partial differential equations*, Trans. Amer. Math. Soc. **90**, 193–254, 1959.

Pillage, L.T. and Rohrer, R.A., *Asymptotic waveform evaluation for timing analysis*, IEEE Trans.Comp. Aided Design **CAD 9 No 4**, 352–366, 1990.

Prada, G. and Bickart, T.A., *Stability of electrical networks containing distributed RC components*, J. Math. Anal. Appl. **33**, 367–401, 1971.

Preis, D. and Shlager, K., *Interconnect rise time in superconducting integrated circuits*, IEEE Trans. Circ. Syst., CAS **35 No 11**, 1463–1465, 1988.

Protonotarios, E.N. and Wing, O., *Theory of nonuniform RC lines*, Part II: Analytic properties in the time domain, IEEE Trans. Circ. Th., CT-**14**, 13–20, 1967.

Putatunda, R., *AUTODELAY: A second generation automatic delay calculation program for LSI/VLSI chips*, in Proc. IEEE Int. Conf. Comp. Aided Design, 188–190, 1984.

Rubinstein, J., Penfield, P.Jr. and Horowitz, M.A., *Signal delay in RC tree networks*, IEEE Trans. Comp. Aided Design **2**, 202–211, 1983.

Ruehli, A.E. (ed.), "Circuit analysis. Simulation and design," North Holland, Amsterdam, 1987.

Ruehli, A.E. and Ditlow, G.S., *Circuit analysis, logic simulation and design verification for VLSI*, Proc. IEEE **71**, 34–48, 1983.

Sakurai, T., *Approximation of wiring delay in MOSFET LSI*, IEEE J. Solid State Circ., SC **18 No** 4, 418–426, 1983.

Sandberg, I.W., *Some theorems on the dynamic response of transistor network*, Bell System Tech. J. **48**, 35–54, 1969.

Sandberg, I.W., *Theorems on the analysis of nonlinear transistor networks*, Bell Syst. Tech. J. **49**, 95–114, 1970.

Sandberg, I.W. and Willson Jr., A.N., *Some theorems on properties of DC equations of nonlinear networks*, Bell Syst. Tech. J. **48, No** 1, 1–34, 1969 a.

Sandberg, I.W. and Willson Jr., A.N., *Some network-theoretic properties of nonlinear DC transistor networks*, Bell Syst. Tech. J. **48**, 1293–1311, 1969 b.

Saraswat, K.C. and Mohammadi, F., *Effect of scaling of interconnections on the time delay of VLSI circuits*, IEEE Trans. Electr. Dev., ED **29 No** 4, 645–650, 1982.

Sattinger, D.H., *Stability of nonlinear parabolic systems*, J. Math. Anal. Appl. **24**, 241–245, 1968.

Sato, K., *On the generators of non-negative contraction semi-groups in Banach lattices*, J. Math. Soc., Japan **20**, 423–436, 1968.

Schwartz, A.F., "Computer aided design of microelectronic circuits and systems," Acad.Press, London, 1987.

Showalter, R.E. and Snyder, C.H., *A distributed RC network model with dielectric loss*, IEEE Trans. Circ. Syst., CAS **33 No** 7, 707–710, 1986.

Showalter, R.E. and Xu, X., *Convergence of diffusion with concentrating capacity*, J. Math. Anal. Appl., 1990.

Singhal, K. and Vlach, J., *Approximation of nonuniform RC-distributed networks for frequency and time-domain computations*, IEEE Trans. Circ. Th. **CT-19**, 347–354, 1972.

So, H.C., *On the hydrid description of a linear n-port resulting from extraction of arbitrarily specified elements*, IEEE Trans. Circuit Th. **12**, 381–387, 1965.

Taylor, A.E., "Introduction to Functional Analysis," John Wiley, New York, 1958.

Terman, C.J., *Timing simulation for large digital MOS circuits*, in "Advances in Computer Aided Engineering Design," A.L. Sangiovanni-Vincentelli ed., Greenwich, 1985.

Tsao, D. and Chen, C.F., *A fast timing simulator for digital MOS circuits*, IEEE Trans. Comp. Aided Design, CAD **5** No 4, 536–540, 1986.

Wang, P.K.C., *On certain maximum properties of Cauchy problems*, J. Math. Anal. Appl. **24**, 136–145, 1968.

Webb, G.F., *Continuous nonlinear perturbations of linear accretive operators in Banach spaces*, J. Funct. Anal. **10**, 191–203, 1972.

White, J.K. and Sangiovanni-Vincentelli, A.L., "Relaxation Techniques for the Simulation of VLSI Circuits," Kluwer Academic, Norwell, 1986.

Willson Jr., A.N., *On the solution of equations for nonlinear resistive networks*, Bell Syst. Tech. J. **47**, 1755–1773, 1968.

Willson Jr., A.N., *New theorems on the equation of nonlinear dc transistor networks*, Bell Syst. Tech. J. **49**, 1713–1738, 1970.

Willson Jr., A.N. and Wu, J., *Existence criteria for DC solutions of nonlinear networks which involve the independent sources*, IEEE Trans. Circ. Syst. **CAS 31**, 952–959, 1984.

Wohlers, M.R., "Lumped and distributed networks," Academic Press, New York, 1969.

Wyatt, J.L.Jr., *Monotone sensitivity of nonlinear nonuniform RC transmission lines, with applications to timing analysis of digital MOS integrated circuits*, IEEE Trans. Circ. Syst., CAS **32** No 1, 28–33, 1985 a.

Wyatt, J.L.Jr., *Signal delay in RC mesh networks*, IEEE Trans. Circ. Syst., CAS **32**, No 5, 507–510, 1985 b.

Wyatt, J.L.Jr. and Yu, Q., *Signal delay in RC meshes, trees and lines*, Proc. IEEE Int. Conf. Circ. Aided Design (ICCAD–84), 15–17, Santaclara, 1984.

Yosida, K., "Functional analysis," Springer Verlag, Berlin, 1974.

Yuan, H.T., Lin, Y.T. and Chiang, S.Y., *Properties of interconnection on silicon, sapphire and semi-insulating gallium arsenide substrates*, IEEE Trans. El. Dev. **ED-29, No 4**, 639–644, 1982.

Zemanian, A.H., *Infinite electrical networks*, Proc. IEEE **64**, 6–17, 1976.

Zemanian, A.H., *The characteristic-resistance method for grounded semi-infinite grids*, SIAM J. Math. Anal **12**, 115–138, 1981.

Zemanian, A.H., *Nonuniform semi-infinite grounded grids*, SIAM J. Math. Anal **13**, 770–788, 1982.

Zukowski, C.A., *Relaxing bounds for linear RC mesh circuits*, IEEE Trans. Circ. Aided Design, CAD **5 No 2**, 305–312, 1986 a.

Zukowski, C.A., "The Bounding Approach to VLSI Circuit Simulation," Kluwer Acad. Press, 1986 b.

Zurada, J.M. and Liu, T., *Equivalent dominant pole approximation of capacitively loaded VLSI interconnection*, IEEE Trans. Circ. Syst., CAS **34 No 2**, 205–207, 1987.